HESBAN

Series Editors

Lawrence T. Geraty
Øystein Sakala LaBianca

ANDREWS UNIVERSITY PRESS  BERRIEN SPRINGS, MICHIGAN

ENVIRONMENTAL FOUNDATIONS:

STUDIES OF CLIMATICAL, GEOLOGICAL, HYDROLOGICAL, AND PHYTOLOGICAL CONDITIONS IN HESBAN AND VICINITY

Contributors

Patricia Crawford
Kevin Ferguson
Dennis Gilliland
Tim Hudson
Øystein Sakala LaBianca
Larry Lacelle

Volume Editors
Øystein Sakala LaBianca
Larry Lacelle

Associate Editor
Lorita E. Hubbard

Assistant Editors
Lori A. Haynes
Sandra L. Penley

Technical Assistants
James K. Brower
Oscar Canale
Tung Isaiah Duong

published with the financial assistance of the
National Endowment for the Humanities
and Andrews University

Andrews University Press
Berrien Springs, MI 49104

© 1986
All rights reserved. Published December 1986.
Printed in the United States of America.

92 91 90 89 88 87 86   7 6 5 4 3 2 1

ISBN 0-943872-15-4
Library of Congress catalog card number 86-072952

Table of Contents

Chapter One **Introduction**   3
    Øystein Sakala LaBianca

Chapter Two **Climate of Tell Hesban and Area**   7
    Kevin Ferguson and Tim Hudson

Introduction 9
Aspects of Paleoclimatology in Hesban and Vicinity 9
Introduction to Macroprocesses 11
    Summer Air Masses and Movements 11
    Winter Air Masses and Movements 13
    Transitional Seasons 13
Methodological Approach to Climate Classification for Hesban 15
    Temperatures at Hesban 16
    Precipitation at Hesban 18
    Condensation at Hesban 18
Climate and Humans at Hesban 21

Chapter Three **Bedrock, Surficial Geology, and Soils**   23
    Larry Lacelle

Introduction 25
Bedrock Geology 25
    Bedrock Characteristics 28
    Influences of Bedrock Geology on Man 31
    Influences of Man on Bedrock Geology 35
Surficial Geology 35
    Physiographic Subdivisions and Landforms 35
    Geomorphic Processes and Physical Characteristics of the Surficial Materials 38
    Influences of Surficial Geology on Man 45
    Influences of Man on the Surficial Geology 45
Soils 45
    Soil Classification 45
    Soils Genesis 48
    Soils Physical Properties 53
    Soil Chemical Properties 53
    Influences of Soils on Man 57
    Influences of Man on Soils 57

Chapter Four **Surface and Groundwater Resources of Tell Hesban and Area, Jordan**   59
    Larry Lacelle

Introduction 61
Surface Water 61
    Influences of Surface Water Resources on Man 69

    Influences of Man on the Surface Water Resources 69
Groundwater 69
    Influences of the Groundwater Resources on Man 72
    Influences of Man on the Groundwater Resources 72

Chapter Five **Flora of Tell Hesban and Area, Jordan**   75
   Patricia Crawford

Introduction 77
Economic Uses of Plants 79
    Plants Utilized as Food 79
    Plants Utilized for Industrial Purposes 80
    Plants Utilized for Domestic Purposes 80
    Plants Utilized for Fodder and Forage 80
Species Descriptions 82
Cultivated Plants of Tell Hesban Area 97
Summary 98

Chapter Six **Ecology of the Flora of Tell Hesban and Area, Jordan**   99
   Larry Lacelle

Introduction 101
General Characteristics of the Flora 101
    Vegetation Zonation and Potential Climax Species 101
    Present-day Flora 105
    Environmental Parameters and Their Effects Upon the Flora 105
Ecological Units of the Tell Hesban Area 110
    Dry-Farmed, Cultivated Fields 110
    Dry, Barren Hillsides 113
    Moist Wadi Floors 113
    Irrigated, Cultivated Fields 113
Summary 119

Chapter Seven **Paleoethnobotany and Paleoenvironment**   123
   Dennis R. Gilliland

Introduction 123
Materials and Methods 124
Results 125
Discussion 125
    Comparison with the Previous Study 125
    Ethnobotanical and Paleoecological Interpretations 125
    Paleoenvironment 139
Conclusions 139

Chapter Eight **Conclusion**   143
   Øystein Sakala LaBianca and Larry Lacelle

Index   147

List of Figures

Fig. 1.1	Location of Hesban, Jordan	4
Fig. 2.1	Paleoclimatic data for the upper Jordan Valley (after Horowitz, 1968)	10
Fig. 2.2	Summer air masses and their movements	12
Fig. 2.3	Winter air masses and their movements	14
Fig. 2.4	Thirty year average rainfall	19
Fig. 3.1	Bedrock Geology of the Tell Hesban area (after Bender 1975)	26
Fig. 3.2	Geological cross-section near Tell Hesban (after Agrar und Hydrotechnik 1977, Map HG 1.3)	27
Fig. 3.3	Physiographic subdivisions of the Tell Hesban project area (after Bender 1975)	37
Fig. 3.4	Cross section depicting surficial materials on the surface of the Transjordanian Plateau	40
Fig. 3.5	Cross section depicting surficial materials in the wadis west of Tell Hesban	42
Fig. 3.6	Cross section depicting surficial materials in the wadis near the floor of the Jordan River Valley west of Tell Hesban	43
Fig. 3.7	Soils of the Tell Hesban project area	46
Fig. 3.8	Profile of a typical Red Mediterranean Soil	47
Fig. 3.9	Profile of a typical Yellow Mediterranean Soil	49
Fig. 3.10	Profile of a typical Yellow Soil	51
Fig. 4.1	Major drainages and locations of springs in the Tell Hesban project area	62
Fig. 4.2	Geological cross section near Tell Hesban (after Agrar und Hydrotechnik 1977, Map HG 1.3)	71
Fig. 5.1	The Tell Hesban project area	78
Fig. 6.1	Ecological units of the Tell Hesban project area	102
Fig. 6.2	Vegetation zonation, Tell Hesban project area	103
Fig. 6.3	Cross section depicting elevational relationships of climate zonation, geology, surficial materials, soils, and vegetation in the Tell Hesban project area	104
Fig. 6.4	Cross section depicting relationships between topography, surficial materials, and vegetation in the Dry Farmed, Cultivated Fields ecological unit	112
Fig. 6.5	Cross section depicting relationships between topography, surficial materials and plant species in the Dry, Barren Hillsides ecological units	115
Fig. 6.6	Cross section depicting relationships between topography, surficial materials and plants species in the Moist Wadi Floor ecological unit	116
Fig. 6.7	Cross section depicting relationships between topography, surficial materials, and plant species in the Irrigated, Cultivated Fields ecological units	118
Fig. 7.1	Graph depicting the number of seeds recovered from each archaeological period for each cultivated or potentially cultivated species and from each of the noncultivated species	126-127

List of Plates

Plate 3.1	Nari on carbonate bedrock in the Wadi Majarr	29
Plate 3.2	Microphotograph of a fossiliferous limestone	29
Plate 3.3	Microphotograph of a very hard millstone from the surface flood on the north slope of Tell Hesban	30
Plate 3.4	Lithic outcrop pattern on the hill across the Wadi Majarr immediately northwest of Tell Hesban	30
Plate 3.5	Tell Hesban from the northwest	32
Plate 3.6	Polished section of fossiliferous limestone	32
Plate 3.7	Microphotograph of fossiliferous limestone	33
Plate 3.8	Lower Cretaceous sandstone in a cave, Wadi Nusaiyat	33
Plate 3.9	Stone block quarry on hill northwest of Tell Hesban	34
Plate 3.10	Partially quarried columnar section	34
Plate 3.11	Grinding wheel of hard fossiliferous limestone	36
Plate 3.12	Microphotograph of a fragment of a basalt bowl	36
Plate 3.13	Microphotograph of a fragment of a basalt bowl	39
Plate 3.14	Topography of the Transjordanian Plateau	39
Plate 3.15	Deep, steep walled wadis west of Tell Hesban	41
Plate 3.16	Fluvial sediments in a wadi below Tell Hesban	41
Plate 3.17	Colluvium mantling lower slopes of a hill near Tell Hesban	44
Plate 3.18	Barren hillsides in the wadis west of Tell Hesban	44
Plate 3.19	Red Mediterranean Soils on the surface of the Transjordanian Plateau Hesban	50
Plate 3.20	Yellow Mediterranean Soils on hilly topography	50
Plate 3.21	Shallow young soils on *nari*	52
Plate 3.22	Profile of Tell Hesban soils	52
Plate 3.23	Ploughing a Red Mediterranean Soil	56
Plate 4.1	Large ancient reservoir at Tell Hesban	63
Plate 4.2	Cistern in depression with channels to collect overland flow during rainstorms	63
Plate 4.3	Remains of an old dam near Ain Hesban	65
Plate 4.4	Illustration of water hauling by donkey, representative of ancient methods used at Tell Hesban	66
Plate 4.5	General view of an area irrigated by water systems from a perennial spring	68
Plate 6.1	Reforested hilltops	106
Plate 6.2	Typical spiny, unpalatable plant species adapted to prolonged grazing pressure	106
Plate 6.3	Denser and more varied vegetation on wadi floors	109
Plate 6.4	Utilization of areas of shallow stony soils for cereal crops	109
Plate 6.5	View of numerous goats and sheep overgrazing cultivated fields	111
Plate 6.6	Goat grazing unpalatable thistles	111
Plate 6.7	General view of plateau surface showing that it is nearly all cultivated with very little area left for natural vegetation	114
Plate 6.8	Shallow stony soils and dissected topography near the summits of hills typical of Dry Barren Hillsides ecological units	114
Plate 6.9	General view of lusher vegetation typical of Moist Wadi Floor ecological units	117
Plate 7.1	Microphotographs of seeds recovered at Tell Hesban	129
Plate 7.2	Microphotographs of seeds recovered at Tell Hesban	130
Plate 7.3	Microphotographs of seeds recovered at Tell Hesban	132
Plate 7.4	Microphotographs of seeds recovered at Tell Hesban	137
Plate 7.5	Microphotographs of seeds recovered at Tell Hesban	138

List of Tables

Table 2.1	Average mean daily temperature in °C by month (30 year average)	17
Table 2.2	Average precipitation in mm by month (30 year average)	20
Table 3.1	Soil physical properties	54
Table 3.2	Soil chemical properties	55
Table 4.1	Irrigated acreage in the Tell Hesban project area	65
Table 4.2	Maximum and minimum flows for major springs in the Tell Hesban project area (after Hydrology Division 1966)	67
Table 5.1	Alphabetical List of the Flora of Tell Hesban and Area	81
Table 7.1	A Systematic List of carbonized seeds identified from Tell Hesban, Jordan	128
Table 7.2	The number of seeds per taxon per sample for each period	134-136

Chapter One
INTRODUCTION

Øystein Sakala LaBianca

Chapter One

Introduction

Studies of various aspects of the natural environment of Tell Hesban, Jordan date back to the second season of excavations in the summer of 1971. In response to an invitation from Siegfried H. Horn, the organizer and director of the first three field seasons at Tell Hesban, Reuben G. Bullard, University of Cincinnati, carried out a geological study of the Hesban region which culminated in the first publication specifically devoted to an aspect of the natural environment of that site (Bullard 1972). At the end of the same season, identifications of numerous bones of wild animals and birds excavated at Tell Hesban in 1971 were made by Johannes Lepiksaar of the Museum of Natural History in Gothenburg, Sweden. This made possible the first preliminary report dealing with the ancient animal life of the Tell Hesban region (LaBianca 1973).

Over the subsequent four seasons, investigations of the natural environment were expanded, thanks especially to the leadership of Lawrence T. Geraty, director of the 1974 and 1976 field seasons, and Roger S. Boraas, the excavations' chief archaeologist. As a result, studies dealing with various aspects of the present-day and ancient plant life (Crawford and LaBianca 1976; Crawford, LaBianca, and Stewart 1976) and animal life (LaBianca and LaBianca 1975; Crawford 1976; Boessneck 1977; Alomia 1978; Boessneck and von den Driesch 1978; Labianca 1986) were published.

The present volume is devoted primarily to a synthesis of more recent findings pertaining to the physical and botanical environment of Tell Hesban and the region which lies approximately within a 10-km radius of this site (fig. 1). This region corresponds approximately to the area surveyed for archaeological artifacts by the Hesban Regional Survey (Ibach 1978). It includes portions of three major physiographic subdivisions, namely a portion of the Transjordanian Plateau, a portion of the highlands at the edge of the Jordan River Valley, and a portion of the wadi system that makes up the east side of the Jordan River Valley.

As explained elsewhere (LaBianca 1978), the impetus for the research which led to each of the contributions presented here was the conviction that inferences about the state of environmental conditions in antiquity must be informed by a thorough understanding of the present-day environment. Thus, in different ways, the studies included here focus attention on specific ecological processes in evidence in Hesban's environment today. The insights thus gained are then used to generate propositions about the environmental conditions which existed in earlier times.

Kevin Ferguson and Tim Hudson's chapter on the climate of the Tell Hesban area presents an overview of the macroclimatic processes which account for the annual fluctuations in temperatures, rainfall, and condensation in the vicinity of this site. While neither of these two investigators participated as field workers in Jordan, their professional expertise as geographers was solicited in the production of this chapter in order that the limited climatological data gathered in the field could be presented in a broader, more informative context. The report which they offer here represents a remarkable achievement, given the meagerness of the data with which they had to work.

Much of the fieldwork which went into the production of Chapters Three, Four, Five, and Six, was carried out in the summer of 1979 by Larry Lacelle, Patricia Crawford, and Hal James, thanks to a grant from the National Endowment for the Humanities. In addition to field observations of plant species, geology, surficial geology, soils, surface and groundwater resources, cultivation, grazing practices, and human settlement patterns, this team sought out maps and technical reports dealing with environmental aspects of the Hesban region from government agencies in Amman. In this, and in many other ways, their work was aided by the kind assistance of Dr. Adnan Hadidi, director general, Jordan Department of Antiquities, and his staff members Nadia Razmuzi and Samir Gishan.

4 ENVIRONMENTAL FOUNDATIONS

Fig. 1.1 Location of Hesban, Jordan.

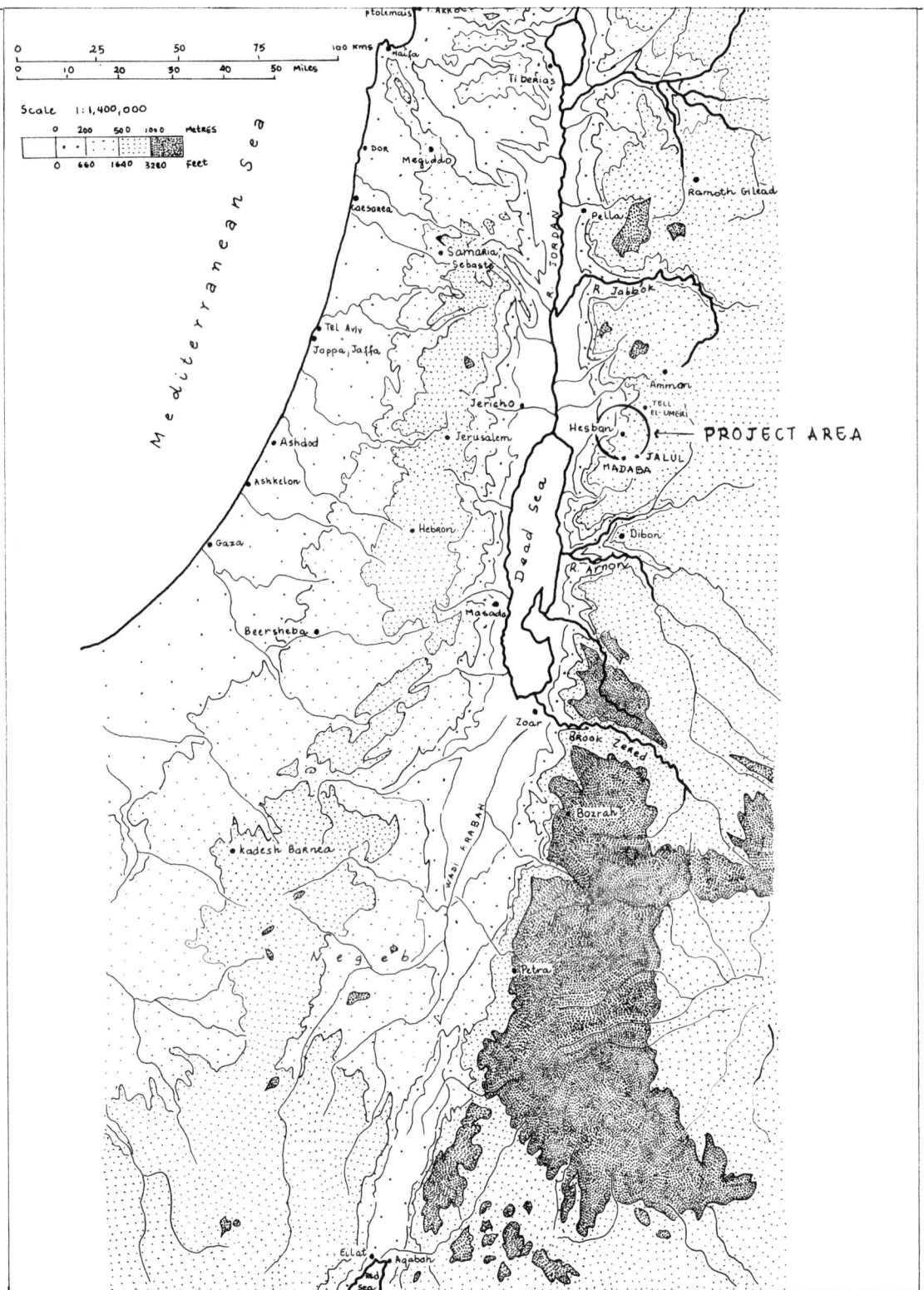

As a soil scientist and landuse analyst with the government of Canada, Lacelle's services were solicited to lend professional support to the analysis of the ecological processes which have cooperated in shaping the landscape in the vicinity of Hesban in the present and in the past. Lacelle's three chapters synthesize his own observations and those of his team members in the summer of 1979, those of his team members, and those of previous investigators of the Hesban region. By examining both the influence of the natural environment on human activity, and the influence of human activity on the natural environment, Lacelle illuminates many environmental and certain cultural changes which have occurred as a result of intensifying human intervention and management of the local landscape over the centuries.

Patricia Crawford's chapter offers descriptions of the habitats and uses of nearly one hundred different plant specimens collected in the vicinity of Hesban during the summers of 1974, 1976, and 1979. Although by no means representative of the entire range of flora which occur in this region throughout the entire year, this analysis of plants which may be found during the dry months of July and August heightens awareness and understanding of the wide range of ecological niches which exist within the project area today. It also serves to emphasis the important contribution which these many different plants make, either directly or indirectly (as fodder for domesticated animals), to the lives of people. Naturally, the investigations which led to this assemblage were a crucial components of the paleoethnobotanical work carried out by Patricia Crawford carried out during the 1976 season at Tell Hesban.

The paleoethnobotanical report by Denis Gilliland presents his analysis of the carbonized seeds collected during the 1978 season of excavations at Tell Hesban. What is published here is a slightly abbreviated version of the Master's thesis completed in 1979 under the direction of Professor Lanny Fisk of the Biology Department at Walla Walla College (College Place, WA). Gilliland's report discloses more than simply the identification of the species collected in the samples from Hesban, but also endeavors to evaluate the carbonized seeds from the perspective of what light they might shed on the extent significant climatic changes have occurred in Hesban's landscape since antiquity. It thus alerts to the contribution which future paleoethnobotanical analysis can make to our understanding of such changes.

In the concluding chapter, a brief discussion is offered regarding the implications of these various reports for understanding Hesban's natural environment since early historical times. While it might be argued that such statements are premature, given the data offered here, our view is that sufficient work has been done to warrant setting forth proposals which might help in focusing future research on the ancient environment in this region. Indeed, two future volumes in this series, namely one dealing with the animal bone finds and another dealing with the human food system, will provide independent lines of evidence against which to evaluate these proposals. In this manner, it is anticipated that the initial perception of the ancient environment offered here will be refined and modified as additional pieces of the puzzle are fitted together.

References

Alomia, M. K.
1978   Tell Hesban 1976: Notes on the Present Avifauna of Tell Ḥesbân. *Andrews University Seminary Studies* 16: 289-303.

Boessneck, J.
1977   Erste Ergebnisse unserer Bestimmungsarbeit an den Tierknochenfunden vom Tell Hesban, Jordanien. *Archaologie und Naturwissenschaften* 2: 55-72.

Boessneck, J., and von den Driesch, A.
1978   Tell Hesban 1976: Preliminary Analysis of the Animal Bones from Tell Hesban. *Andrews University Seminary Studies* 16: 259-288.

Bullard, R. G.
1972   Geological Study of the Heshbon Area. *Andrews University Seminary Studies* 10: 129-141.

Crawford, P.
1976   Tell Hesban 1974: The Mollusca of Tell Hesban. *Andrews University Seminary Studies* 14: 171-176.

Crawford, P., and LaBianca, Ø. S.
　1976　The Flora of Hesban. *Andrews University Seminary Studies* 14: 177-184.

Crawford, P.; LaBianca, Ø. S.; and Stewart, R.
　1976　The Flotation Remains. *Andrews University Seminary Studies* 14: 185-188.

Ibach, R.
　1978　Tell Hesban 1976: Expanded Archaeological Survey of the Hesban Region. *Andrews University Seminary Studies* 16: 201-214.

LaBianca, Ø. S.
　1973　Tell Hesban 1971: The Zooarchaeological Remains. *Andrews University Seminary Studies* 11: 133-144.

　1978　Man, Animals and Habitat at Hesban: An Integrated Overview. *Andrews University Seminary Studies* 14: 229-252.

　1986　The Diachronic Study of Animal Exploitation at Hesban. Pp. 167-182 in *The Archaeology of Jordan and Other Studies: Presented to Professor Siegfried H. Horn*, eds. L. T. Geraty and L. G. Herr. Berrien Springs, MI: Andrews University Press.

LaBianca, Ø. S., and LaBianca, A. S.
　1975　Tell Hesban 1973: The Anthropological Work. *Andrews University Seminary Studies* 13: 235-247.

Chapter Two
CLIMATE OF TELL HESBAN AND AREA

Kevin Ferguson
Tim Hudson

Chapter Two
Climate of Tell Hesban and Area

Introduction

The impact of climate on human activity, past and present, can best be explained in relation to a normative of factors related to climate conditions at Hesban, Jordan. This involves discussion of climatic processes which affect the region today and operated in the past, description of pertinent aspects of the paleoclimate, an examination of macro and microclimate processes, and classification of Hesban's climatic regime based on available data and in the context of standard classification procedures.

Before any detailed synopsis of macroclimatic processes and their characteristics is attempted, the limitations of the study must be introduced. The most serious limitation is the lack of data. As in previous climatological studies of Jordan (Manners 1969), consideration must be given to the reliability and adequacy of collected data. In Jordan, these two factors are limited by the absence of long-term climatic data, the unequal distribution of meteorological stations, and incomplete records. The quality of available data is poor. Yet these data provide the only basis for an analysis such as this. Attempts are underway to upgrade the meteorological observation capabilities in the Hashemite Kingdom of Jordan. In the near future, the prospect is for more observations of finer quality and greater accuracy. That the majority of stations now operating in Jordan have been established comparatively recently indicates a commitment on the part of Jordan to assess its climatic foundation from a proper data base. Such an assessment is essential for proper understanding of the human/environment interface within Jordan.

In 1976 Robin Cox, Andrews University (Berrien Springs, MI), began to record basic weather parameters at Tell Hesban through the use of equipment garnered under the auspices of Prince Ra'ad of Jordan, and Ghazi El-Risai, director of the Department of Antiquities. While these observations were short-lived, they have helped stimulate an ongoing interest in the macro as well as the microclimatic processes operating in the area, an interest initiated by anthropologist Øystein LaBianca (LaBianca 1978: 237).

Aspects of Paleoclimatology in Hesban and Vicinity

Considerable difficulties surround attempts to determine the exact paleoclimatic conditions in Ḥesbân. Through a generalized analysis of young sediments and pollen, Horwitz (1968) outlines the basic trends in climatic change for the Upper Jordan Valley over the last 60,000 years. A study of this sort is open to numerous interpretations. Nevertheless, the general trends uncovered and articulated are in consonance with those found by researchers elsewhere (Kubiena 1962).

Although the Horwitz study illustrates conditions located within 160 km of Hesban and not at Hesban itself, some assumptions can be made linking paleoclimatic trends for both areas. In fact, local variations in the climate of this region have likely increased over time, making the historical analogy more tenable. The evidence suggests that periodically cooler, moister climatic conditions existed in the vicinity of Hesban prior to the Holocene (recent) geological epoch. Periods of fluctuating temperatures associated with rather constant, high humidity levels were characteristic of the Pleistocene epoch. With the Holocene came a gradual rise in temperature norms (fig. 2.1). Humidity levels, however, while decreasing generally, display marked irregularity fluctuating between periods of unusual dryness and water surplus. In recent history it is likely that a macrotrend towards slightly higher atmospheric moisture levels exists. This factor remains speculative, and local

10 ENVIRONMENTAL FOUNDATIONS

Fig. 2.1 Paleoclimatic data for the upper Jordan Valley (after Horowitz, 1968)

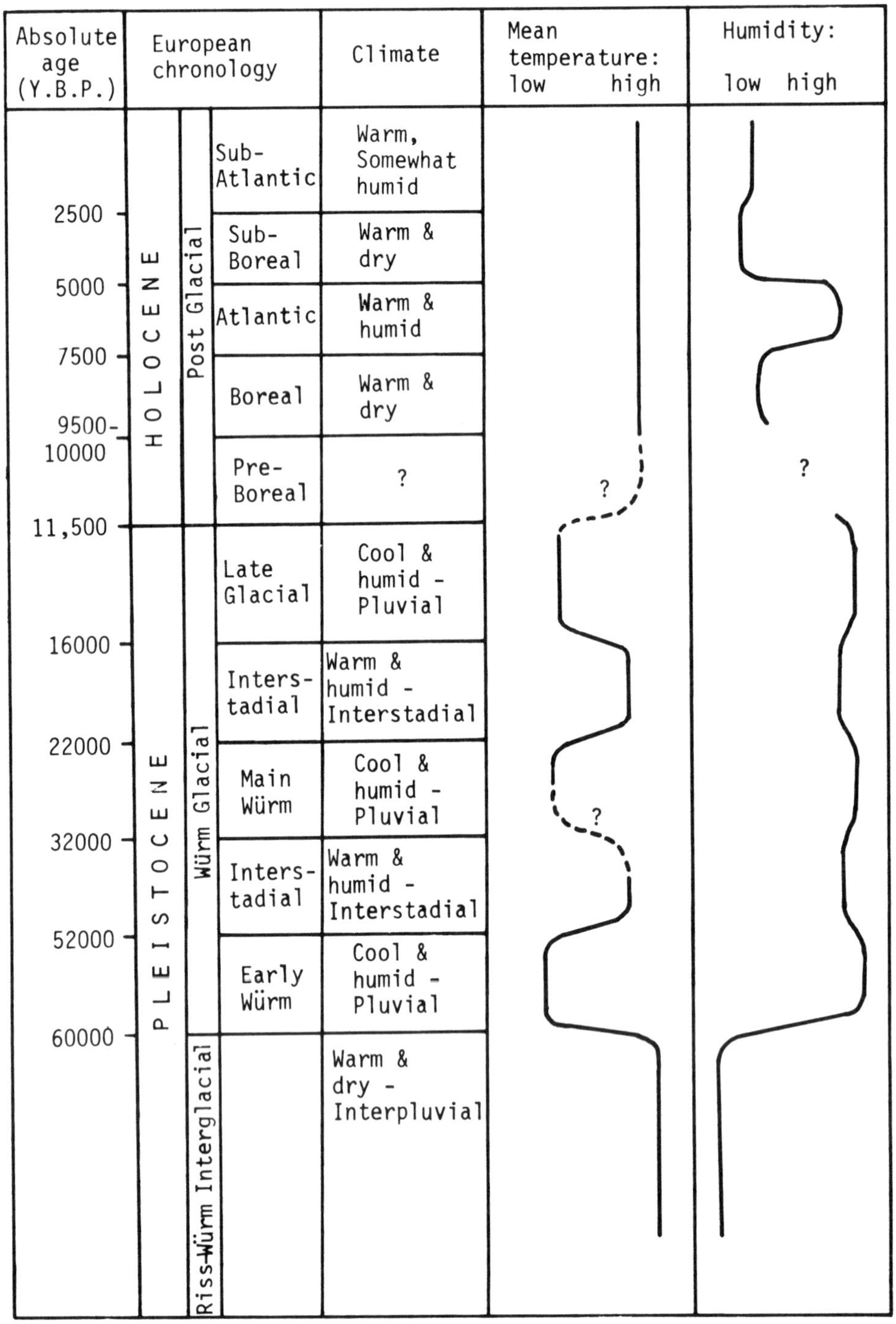

Absolute age (Y.B.P.)	European chronology			Climate	Mean temperature: low high	Humidity: low high
2500	HOLOCENE	Post Glacial	Sub-Atlantic	Warm, Somewhat humid		
5000			Sub-Boreal	Warm & dry		
7500			Atlantic	Warm & humid		
9500–10000			Boreal	Warm & dry		
11,500			Pre-Boreal	?	?	?
16000	PLEISTOCENE	Würm Glacial	Late Glacial	Cool & humid – Pluvial		
22000			Interstadial	Warm & humid – Interstadial		
32000			Main Würm	Cool & humid – Pluvial	?	
52000			Interstadial	Warm & humid – Interstadial		
60000			Early Würm	Cool & humid – Pluvial		
		Riss-Würm Interglacial		Warm & dry – Interpluvial		

empirical evidence to support it has not been forthcoming.

Along with climatic change, changes in landuse and agricultural practices likely occurred. These changes left their impression on Hesban's landscape. The influence of contemporary vegetation cover on climate is minimal. Transpiration of moisture into the atmosphere at Hesban is also of little consequence and has only a nominal effect on the local climate. In a reverse relationship, however, climate does limit vegetative growth by requiring plants to develop largely beneath the soil surface rather than above it.

These paleoclimatic conditions are the backdrop against which historical macroclimatic weather patterns act in establishing life parameters for the Hesban region. These macroprocesses function in relation to the paleoconditions touched on above and, in one sense, represent temporal aberrations in epochal patterns.

Introduction to Macroprocesses

Jordan's location—inland from the Mediterranean Sea and on the edge of one of the great desert belts of the earth—forms the basis for the character of its climate. Actual climatic conditions in Hesban, Jordan and vicinity are results of dynamic forces of air that cover a much larger part of the earth's surface and atmosphere, and in which local differences are only variations. The situation of Jordan between 33° N and 29° N latitude defines only two climatic seasons—a slightly rainy season from November to March, and a very warm, dry-weather season for the rest of the year. Generally, elevation and relief for the greater part of the East Bank of Jordan introduce local variations in annual temperature and precipitation regimes. The microclimate of Hesban, which is located on the East Bank approximately 880 m above sea level, exemplifies this variable effect. Potential evapotranspiration exceeds the total amount of precipitation during the summer months, creating a deficit in the water balance budget for Hesban. From November to March, most precipitation occurs in conjunction with the period of low evaporation; thus, Hesban and vicinity enjoy a meager surplus in the water budget. Manners (1969) found that despite an increase in rainfall during the winter months, the soil water is basically deficient in terms of agricultural needs, a condition which characterizes the Lower Jordan Basin.

Summer Air Masses and Movements

In summer, a thermodynamic high pressure, which develops over the thermic equator and then shifts north over the earth's equator north of 40° N latitude, governs the upper air masses over Hesban and vicinity. The upper air and its associated high pressure prevents the formation of strong depressions and the convection of surface air masses, and thus causes a thermic inversion over the Mediterranean Sea. This inversion, in which lower layers of air are cooler than the air at higher altitudes, generates stability in the air and often impedes the formation of precipitation-bearing clouds, even if relative humidity reaches 85%. This condition is not uncommon during the summer months in the vicinity of large expanses of water-deficient land.

Surface air masses that enter the Jordan region in summer are of tropical origin, due to the equatorial low over East Africa which moves as far north as 25° N and merges with the monsoonal lows of Asia. The subtropical high pressure cells of North Africa disintegrate, owing to high surface temperatures with the main cell of subtropical pressure now located over the Azores in the Atlantic Ocean (fig. 2.2). The monsoonal low of Asia, which generally lies over the Indus plain and Baluchistan, continues through southern Iran to the Tigris-Euphrates lowlands, while a secondary low usually develops over Cyprus. Thus, a steady air movement occurs toward the east and southeast as a result of the pressure gradient existing from the Atlantic to the Persian Gulf. Some alteration of this general summer airflow may occur as the Cyprus low air flows around the existing pressure gradients in an opposite, or counter clockwise, direction. Variable wind directions may result from this activity.

The summer season's most significant feature is the constant atmospheric condition. Average temperatures for a single month usually correspond closely to actual daily temperatures. The difference of average monthly temperatures May through September is small. Skies are usually free of clouds and rain occurs only on rare occasions—and then only as local showers. These showers are generally short in duration and

Fig. 2.2 Summer air masses and their movements

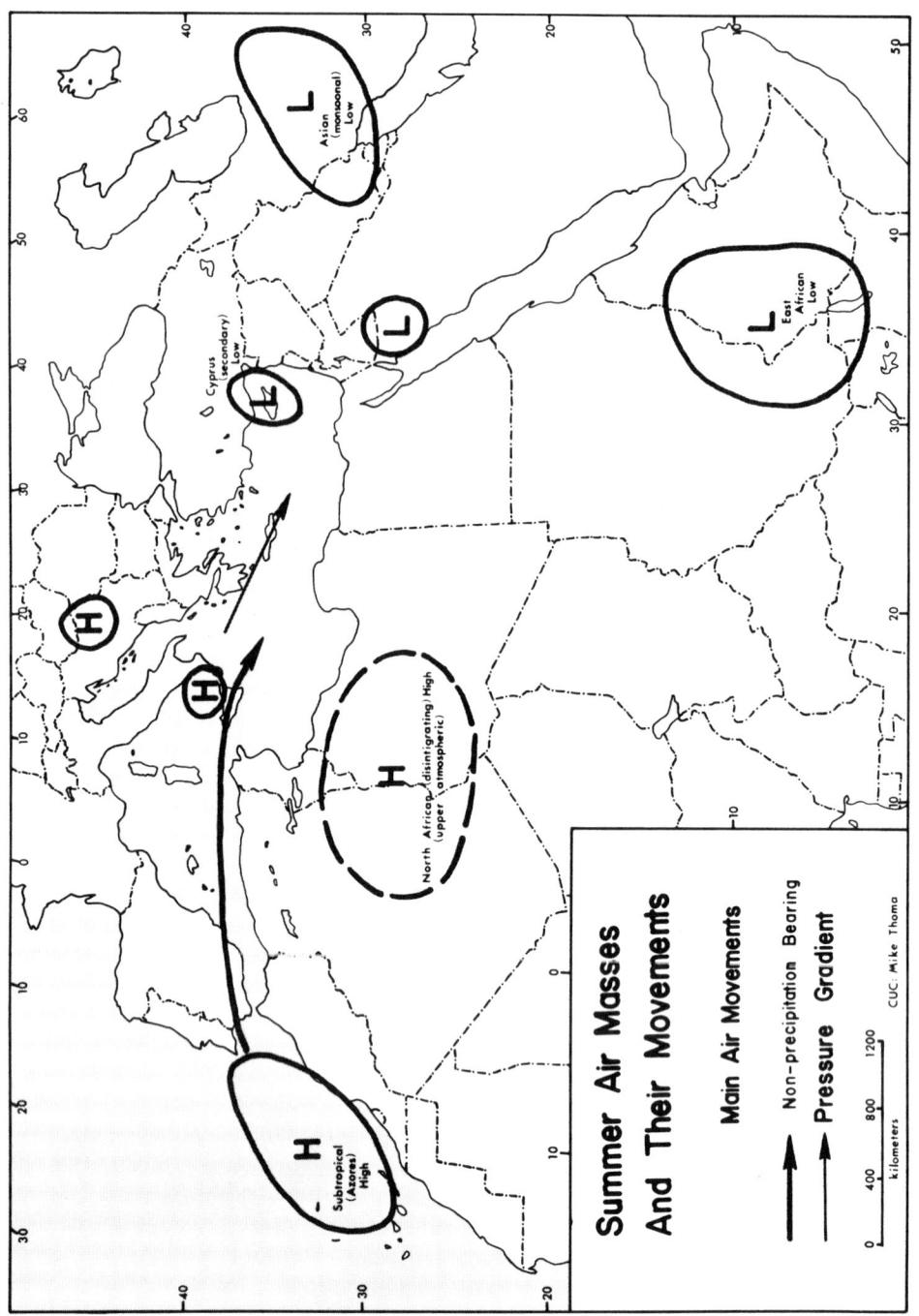

provide only nominal precipitation. These relatively uniform atmospheric pressures and conditions during the summer months show a marked difference from the succeeding low pressure cells and cold fronts that characterize winter months. A shift in these air masses could lead to variations in the weather patterns for a given locale, depending on the nature of the macroclimatic shift. Weather events occurring from these aberrations are difficult to assess in both a contemporary and an historical sense.

Winter Air Masses and Movements

In winter the movement of upper air over the Middle East is accelerated. Causal factors are related to a higher gradient of the thermodynamic pressure in the northern hemisphere because of extreme cooling of the stratosphere during the long polar winter. In addition, migration of the jet stream to points south of the Himalayan Range influences airflow patterns in the Middle East. Karmon (1971) indicates that the westerly jet stream at 100 millibars generally passes over the area at about 35° N latitude, but its path is meandering and southward deflections of the stream can bring depressions with cold air aloft as far south as 29° N latitude.

The Central Asian High dominates surface air in Jordan, Asia, and large portions of Europe during the winter months. Ridges of this barometric high bring cold, dry air to Jordan from two directions. Cold air descends directly from the mountains of Iran over the Syrian Plateau and eastward, or it enters the area as ridges of high pressure from the Balkans.

Because of lower surface temperatures in winter over North Africa, a cell of constant high pressure forms over Libya. Between the Libyan High pressure cell and the subtropical high over North Africa, a low pressure trough develops over the Mediterranean (fig. 2.3). This development is linked with the heat reservoir capabilities of the Mediterranean Sea—capabilities which emerge from the differential heating capacities of land and water. This winter low over the Mediterranean is usually observed as the forward edge of the polar air masses the Polar Front. High pressure belts both to the north and to the south, funnel into the trough, along with Atlantic depressions, bringing with them cyclonic development and rainfall. These depressions, and their associated air movements are the main source of rains over Jordan—and Hesban—throughout the winter.

Basically, then, there are four kinds of air masses influencing weather in Hesban and vicinity during the winter months:
1. warmed, moisture-laden polar air which forms the center of the Mediterranean depressions and is the major source of rainfall;
2. tropical air from the south and southeast responsible for dry, warm weather and also noted for accompanying dust storms which sometimes evolve into short-lived rainstorms;
3. polar air from the Asian High flowing directly into Jordan. Frost during the night is frequently associated with the invasion of cold, dry air, as is clear weather; and
4. polar air from the Balkans or southern Russia, flowing in the wake of depressions and bringing either cold, moist air from the Mediterranean or cold, dry air from southern Russia. These intrusions are also short-lived.

The winter months are marked by irregular and unstable weather patterns. Diverse seasonal, annual, and cyclical averages very rarely indicate uniformity in climatic conditions. In many instances, even fifty-year averages are not sufficient to establish a successful model for a winter season. Rapid changes in temperatures at recording stations, associated with the invasion of differing air masses, and variability in the amounts of rainfall which may occur from place to place, create great difficulty in obtaining accurate, reliable data. Rainfall amounts for a year may deviate as much as 50% on both sides of the multi-annual average. The reliability of the data base, then, should be considered when classifying any climatic regime. In the case of Jordan, with its particular climatic fluctuations, the nature of the data makes such normative descriptions tenuous, especially for agricultural planning, where rainfall data is critical.

Transitional Seasons

The changing of seasons within Jordan is short and irregular. In April, the Asian High begins to retreat and collapse. However, its disappearance is not completed until mid-May. June indicates the full development of the

14 ENVIRONMENTAL FOUNDATIONS

Fig. 2.3 Winter air masses and their movements

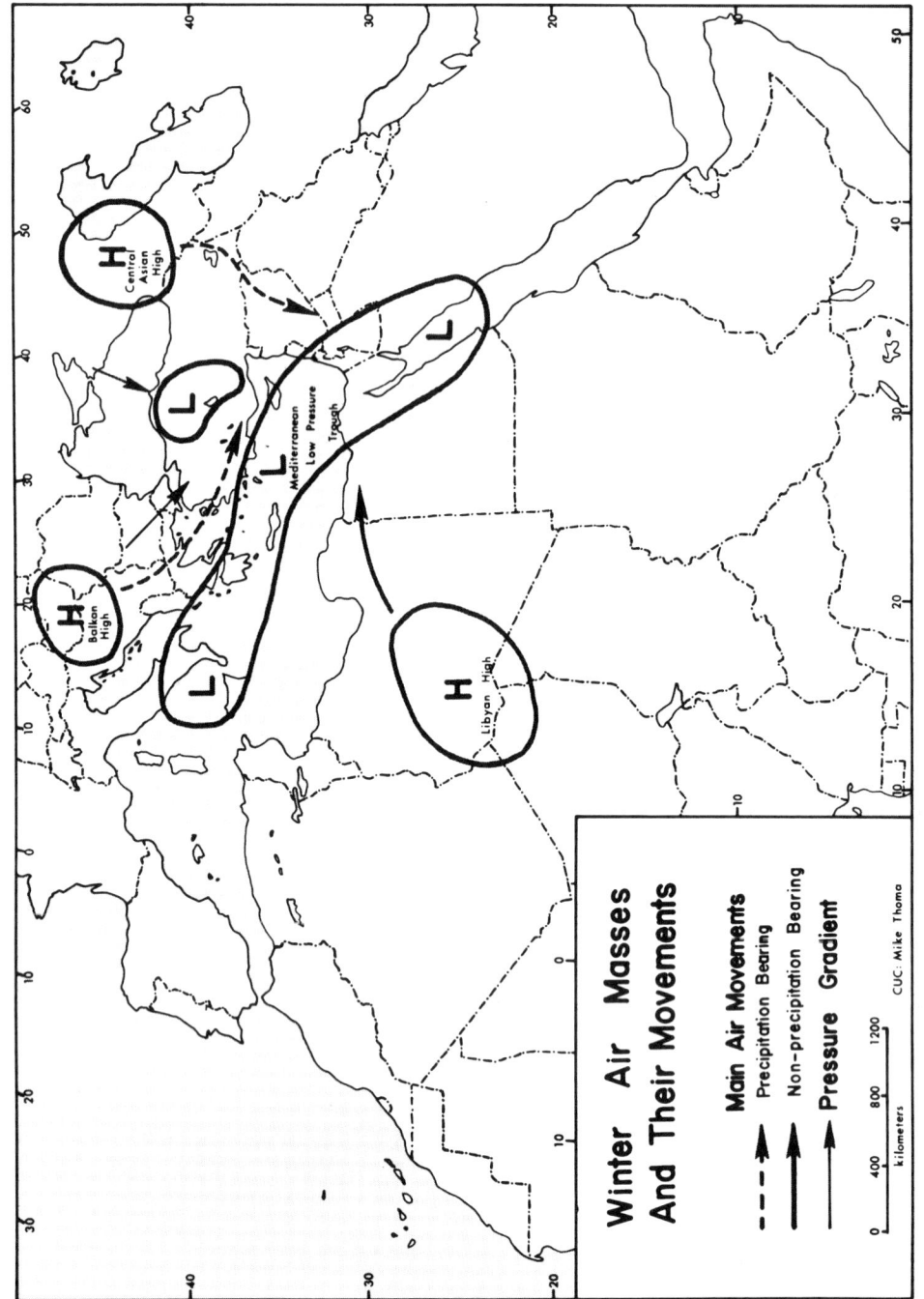

Azores High and the beginning of western air circulation over Jordan.

The *khamsin* depressions, also known as *sirocco*-type winds (very dry and usually accompanied by dust storms), frequent Hesban and vicinity before and after the summer season. This dry wind from the south or southeast sometimes reaches gale force and is heralded by a falling barometer, a drop in relative humidity below 10%, and a hazy sky. Observations indicate that within a few hours a rise in temperature in the neighborhood of 4.4° to 10° C can occur (Hashemite Kingdom 1971). Although the windstorms are often noted for their periodicity (lasting 24 to 72 hours) they also cause considerable discomfort and can have dramatic effects on agriculture; they carry the ability to completely destroy crops.

Another climatological phenomenon related to the transition of seasons is the *shamal*. The *shamal* wind event is generated from the north or northwest generally at intervals between June and September. This wind extends for as long as 9 or 10 days. Usually the *shamal* blows steadily during daytime and slows at night. Both the *sirocco* and *shamal* winds are repetitive. The *shamal*'s dryness is attributed to its origin as a mass of polar air warmed by its passage over the Eurasian landmass and its eventual subsidence. Its dryness allows intense heating at the earth's surface, resulting in high day-time temperatures which decrease only moderately after sunset.

The macroprocesses discussed here set the stage for the local climate at Hesban. Local climates can be viewed as spatial variations in macrosystems. More often than not, however, they illustrate rather than contradict macroprocesses.

Methodological Approach to Climate Classification for Hesban

The precise classification of the Hesban climatic regime is difficult, as it appears to lie in a transitional zone between semiarid and arid regions. Moreover, this marginal locality is, as always, modified by Hesban's relatively high altitude of 880 m. Semiarid climates are generally associated with yearly rainfall totals of 250 mm to 500 mm and with steppe-type vegetation. Arid climates, on the other hand, usually receive less than 200 mm of precipitation per year and are associated with desert vegetation. In both regions, evapotranspiration rates exceed total amounts of precipitation. Hesban's precipitation regime, with approximately 390 mm annually at the tell, is more semiarid than arid, while its vegetation is both steppe and transitional to desert (Chapter Five).

Judging from the location of the 200 mm isohyet seen on the rainfall map of figure 2.4, Tell Ḥesbân is located approximately 25 km north and 25 km west of the precipitation regime associated with desert. In the standard Koeppen classification system (Strahler 1973), Hesban would correspond more closely, to a BSk, a dry steppe with a cold winter. This classification has implications for various climatic variables, including temperature and precipitation. Thermally, a BSk climate is one where the average annual temperature is less than 18° C, and the average temperature of the warmest month exceeds 18° C. The relationship of this heat regime with precipitation amounts of 375 mm and upwards forms the BSk classification which characterizes the humid margins of upland deserts throughout the world.

General information on microclimatic processes active in Hesban and vicinity result from scientific measurements of the following climatic indices: standard time, pressure temperature, vapor pressure and relative humidity, clouds, sunshine duration, wind, evaporation, condensation, and precipitation. These processes generate the microclimate of Hesban. The methods of measuring these processes employed by the Hashemite Kingdom of Jordan, Ministry of Communications Meteorological Department, are as listed:

1. Standard time used in Jordan is the Local Standard Time. LST=GMT + 2 hours.
2. Pressure values are observed to the nearest tenth of a millibar (mb) reduced to 0° C and corrected to normal gravity.
3. Temperature measurements are recorded at 1 meter above the ground surface in Stevenson screen. Both maximum and minimum thermometers are usually observed at 08:00 local time, but the value of the maximum temperature is derived from a comparison with the previous day's reading. Surface minimum thermometers are precisely installed at 5 cm above mostly bare soil and in unshaded area. Soil thermometers are in direct contact with the soil.

4. Values of vapor pressure (mb) and relative humidity (%) are obtained from unventilated wet and dry bulb thermometer readings and calculated according to appropriate standards.
5. Cloud amounts result from visual observations and are registered in degree of sky coverage.
6. A Campbell Stokes recorder is utilized to obtain values for sunshine duration.
7. Wind observations are recorded by instruments such as the Dines pressure tube anemograph, Hand cup anemometers and normal wind vanes. Recorded units of measure are in speeds to the nearest knot.
8. Evaporation rates are monitored by utilizing the Piche evaporimeter.
9. Dew measurements resulting from the Dudevani dew gauge are recorded in millimeter water columns (1 mm=kg/m squared). The gauge is installed at 30 cm and 100 cm above the ground.
10. In most stations, locally made rain gauges are utilized for recording precipitation and are installed at 100 cm above the ground. Snow depth measurements are taken in centimeters. Both maximum daily depth (most recent layer of snow during 24 hours) and maximum total depth (deepest layer of snow accumulated and retained by the ground) are equivocated to water amounts.

These procedures utilized in Jordan are spelled out in order to show the surface technicalities of weather recording and to foster some appreciation for the delicacy and difficulty of this task. These methods are universal.

Although no meteorological observation station is located in Hesban, an estimate of microclimatic conditions can be established by analyzing microclimatic conditions and data from Madaba and Naur, the two nearest stations. An interpolation of measurements obtained from these two stations and corrected by atlas readings (see References), is projected as an estimate of microclimatic conditions at Hesban. Data bases through time for Madaba and Na'ur are limited. Although temperature and precipitation are the main indices for climatic classification, all other indices are functionally related. Wind, condensation, evaporation, relative humidity, pressure, and clouds must be accounted for in terms of these functional relationships to provide an accurate picture of normative weather processes at Ḥesbân.

Other meteorological events, such as frost, hail, dust storms, and thunder and lightning storms which occur at Ḥesbân, can be referred to as special phenomena. The periodicity of these events underscores the difficulty in generally assessing their effects. No effort will be made here to describe the magnitude and frequency of such events at a detailed level due to the scarcity of accurate data concerning these phenomena. The possibility of their occurrence, and of the occurrence of a rare event, however, should be kept in mind. Emphasis here is placed on temperature, precipitation, and condensation, as these climatic phenomena appear to detect behavioral and perceptual responses from local inhabitants. These climatic parameters, then, affect most directly the human/land relationships of Hesban.

Temperatures at Hesban

The mean annual temperature for Hesban is approximately 17° C, with the warmest monthly average, 25° C, occurring in August (table 2.1). The coldest monthly mean temperature is 8° C for January. The hot summers and relatively cold winters are offset by rather mild but short transitional seasons. Temperature extremes as reflected in seasonal variations are important in understanding the intricacies of a microclimate. The difference between the mean daily temperature of the warmest and coldest month for Ḥesbân is 17° C. This is fairly normal for semiarid climates of similar elevation and latitude and for BSk climates in general. The mean daily range of temperature for the winter months, November to March, is 10° C and for the summer months, 14° C. On an annual basis, the mean daily range of temperature (diurnal range) is approximately 13° C. This range is an indicator of the intensity of radiational cooling and heating in summer and winter months, and is linked to the macroprocesses which produce summer stability and winter instability.

Usually submerged within any classification scheme are a number of important details. Averages tend to hide the effects of short-term meteorological events, for example the *khamsin* winds which can trigger a rise in temperature of 4.4° to 10° C in just a few hours. Cold snaps or spells can also occur within the Hesban

Table 2.1 Average mean daily temperature in °C by month (30 year average)

Place	J	F	M	A	M	J	J	A	S	O	N	D	Annual
Madaba	8	8	11	15	19	23	25	25	21	21	15	9	17/16.5*
Na'ur	8	9	11	16	19	23	25	25	21	16	15	9	17/16.3*
Hesban	8	8	12	15	19	23	25	25	21	21*	15	9	16.6**

Source: Climatological Data Annual Summaries (see references); Climatic Atlas of Jordan (see references).

* Data corrected significantly by Atlas reading.

** Data for Hesban is estimated--see text.

region; temperatures drop well below freezing, causing short-lived hoar frost. Frost frequency and magnitude within Hesban and vicinity can only be guessed at, however, as no reliable data is available. December and January are the most likely months in which frosts would occur. Low temperatures during these months may also trigger light snowfall, although snow cover of ground surfaces within the vicinity is short-lived.

Precipitation at Hesban

Weather conditions in general, and precipitation conditions in particular at Hesban are typical of East Bank settlements of similar elevation such as Madaba and Naur. The location of these settlements in the lee of high ground on the West Bank subjects them to frontal precipitation and occasional thunderstorms. Rain of cloudburst proportions are not uncommon during winter and are usually preceded by unseasonably dry weather. Light drizzle also occurs over Hesban and vicinity as a result of the passage of warm fronts associated with the *khamsin* depression (Manners 1969).

As stated earlier, Hesban's mean annual precipitation is estimated at 392 mm. This interpolated amount is suggested as the average, but it should be noted that rainfall throughout the vicinity is exceedingly variable and often reflects local relief or other conditions (fig. 2.4). The greatest amount of precipitation, 89 mm, occurs in January. Nil or trace amounts of precipitation are found in June, July, and August (table 2.2). Winter precipitation at Hesban is usually in the form of rain, sleet, snow, or hail, with rain being overwhelmingly predominant. Precipitation is generally accompanied by cloud cover with variable duration but which may last for days. The importance of winter precipitation in any form cannot be overemphasized in terms of the human interface. Agricultural development is largely controlled by the availability of water. The seasonality associated with Hesban's precipitation establishes certain limitations on agricultural activity and practices. Its winter months are attractive to pastoralists and dry farmers as the intensity of its operative hydrological cycle is at maximum during this time.

Streamflow is relatively nonexistent on an annual basis, but rapid infiltration, storage, and runoff rates succeed in recharging aquifers in the local area. Emergent springs, or groundwater aquifers exposed at the surface by wadi erosion, also offer a source of water to human and animal populations in areas such as Ain Hesban and Ain Musa (Chapter Four). Many of the springs flow only on a seasonal basis, yet they are fundamental to the maintenance of adequate water supplies in the region.

Condensation at Hesban

Another important source of moisture is derived from the condensation processes operating at Hesban. In essentially dry climates like that of Hesban, frequent night dews significantly contribute to the meager water supply available for plant life. Although no data for dew frequency or amounts exist for Hesban, the processes are active. Dew results from radiational cooling which occurs on nights with clear skies and little wind. Trewartha (1968) found that in sub-humid Palestine, the maximum annual dew water amounts to 55 mm. Most of this occurs in summer when water is at a premium. Since plants absorb water vapor through their leaves, this moisture is of considerable biotic value. It is, in fact, the main source of atmospheric water during the summer months.

Another form of surface condensation for Hesban is fog. Although fog is infrequent at Hesban, it also plays a prominent role in sustaining the region's vegetative life and is a factor in agriculture. The principal type of fog for Hesban is radiation fog which builds up by gravitation when air drainage causes the heavier and colder air to collect at ground level. Radiational ground fog has a distinct diurnal periodicity and is usually short-lived. Another variety of radiation which occurs in Hesban and vicinity is known as high-inversion fog. Inversion fog results from warmer, drier air overlying cooler air at the surface; the upward movement of air is suppressed. The continued cooling of surface air for a succession of nights may create foggy conditions. Ideal conditions for high inversion fogs are similar to those for ground fogs except that inversion fogs are caused by cooling which results not from one night's radiation but from a cumulative net loss of heat by radiation over a period of several days. The moist polar air which intrudes and stagnates on occasion over Hesban, provides ideal temperature and moisture conditions for such fogs. The difficulties in assessing the total dew or fog surface coverage

Fig. 2.4 Thirty year average rainfall

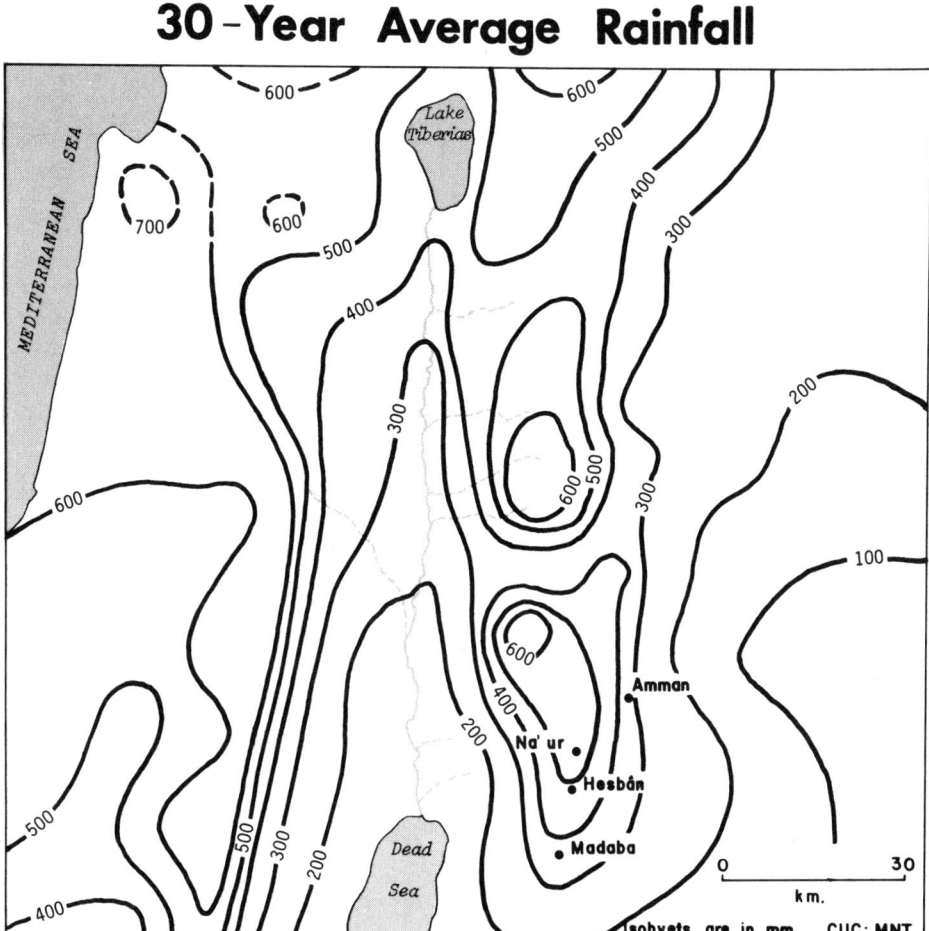

Table 2.2 Average precipatation in mm by month (30 year average).

TABLE 2.2 Average precipitation in mm by month (30 year average)

Place	J	F	M	A	M	J	J	A	S	O	N	D	Annual
Madaba	85	90	70	10	3	-	-	-	2	5	40	75	380/355***
Na'ur	90	99	76	20	10	-	1	1	1	5	35	80	417/523***
Hesban	89	96	70	15	7	-	-	1	3*	5	35	71*	392**

Source: Climatological Data Annual Summaries (see references); Climatic Atlas of Jordan (see references).

* Data corrected significantly by Atlas reading.

** Data for Hesban is estimated--see text.

*** Data taken from Annual Summary estimate; differs from Summary totals and Atlas

and amounts have not yet been completely overcome by contemporary measurement technology. Nonetheless, the realization of the periodical appearance of such phenomena is critical when examining a water-deficient region such as Hesban.

Climate and Humans at Ḥesbân

The entire landscape of Jordan illustrates how the combination forces of humans and climate have modified natural vegetation. Through time, Hesban's successional trends have been interrupted either by meteorologic variations such as droughts or hoar frosts, causing vegetation to shrivel and die, or by humans clearing away native species to till the soil. In general, the intensity of human intrusion has been considerable. The construction of buildings, homes and roads, the advent of grazing, and the need for fuel all represent human-induced alterations in natural ecological systems and make it extremely difficult to assess climatic impact on vegetation.

The natural vegetation cover of Hesban, however, follows a pattern bounded by climate and soils. Chapter Five investigates the flora of Hesban and its relationship to productive technology. It includes a list of Hesban flora that may be found within the region today. Crawford and LaBianca's (1976) assumption that gross climatic conditions in the area have not changed appreciably during the last ten thousand years is correct, although the Horwitz (1968) study indicates that some fluctuations in humidity levels probably have occurred. Moreover, short-term variability in microclimatic conditions can leave an impoverished impression of native flora.

No doubt humankind is the greatest influence on vegetation throughout the world. Plants may adjust to low amounts of precipitation and soil moisture deficiencies by storing nutrients for the dry season in bulbs, tubers, or rhizomes. Still others become even more structurally sophisticated in their adaptation to environment. These adaptations represent long-term adaptations. Similar coping mechanisms are inappropriate for dealing with human-induced ecological changes, which tend to be catastrophic from a biotic perspective.

Ultimately, the removal of native vegetative cover for agricultural purposes creates certain environmental problems which are climatically related. The removal of the vegetation impedes the processes of interception, stemflow, and throughfall. During precipitation events, rain splash is thus maximized on bare soil. Rain splash can initiate rill erosion which is followed by gully erosion with a resultant decrease in soil fertility and arable area. Increased runoff accompanied by wind erosion, due to the lack of ground cover, subjects the fragile soils of Hesban to accelerated climatic processes at an even greater degree. Thus, whatever impact climate has in a normative sense on humans and their livelihoods has been enhanced by human alteration of natural ecological cycles. The result of such alterations is to decrease the available options.

Agricultural activities have no doubt been historically bound by such climatic parameters. Technological innovation has been fostered, many believe, by this boundedness. Change in climatic parameters—when it occurs—is gradual, allowing slow accumulative adaptation. To hypothesize, however, that changes in environment lead to stagnant life modes and therefore dynamic climates lead to experimentation with life mode alternatives, is a great leap of faith. There is little doubt that the climatic bounds elicit a behavioral response in terms of agricultural activity. This response carries, however, many other elements such as the range of technological knowledge and life/group goals and values. Contemporary evidence (Kates and Wohlwill 1966) suggests that behavior is in response to perceived environmental and climatic factors, perceptions which may or may not be closely related to reality. Thus even the establishment of a concrete climatic description does not ensure a causal connection with environmental behavior even where this behavior is believed to be rationally bounded or is completely understood.

While climate may influence human activities, and climatic change may alter normative patterns, climate is more appropriately viewed as one variable among many which establishes the option setting against which humans act. Individual activities within this setting are likely to be idiosyncratic and temporally variable, just as climate itself tends to be. Above all else, however, it is clear that the deficient water balance prevalent in Hesban has shaped life modes to a degree greater than any other environmental/climatic factor.

References

Pertinent materials on climatic processes, on Jordan and on the locality of Hesban are voluminous. For further citations see the excellent bibliography in Manners (1969).

Crawford, P., and LaBianca, Ø. S.
   1976   The Flora of Hesban. *Andrews University Seminary Studies* 14: 177-84.

Hashemite Kingdom of Jordan, Amman.
   1931-60 Central Water Authority. Technical Paper 31.

   1936-63 Technical Papers. Department of Irrigation and Water Power, Nos. 1, 12, 13-20, 20-32.

   1964   Thirty Year Average Rainfall in Jordan.

   1964-76 Climatical Data, Annual Summaries. Meteorological Department.

   1971   *Climatic Atlas of Jordan*. Meteorological Department.

Horwitz, A.
   1968   *The Plinology of Young Sediments in the Upper Jordan Valley*. (From the Hebrew) Jerusalem: The Institute of Oil and Geophysics Research Report No. 1031.

Karmon, Y.
   1971   *Israel: A Regional Geography*. New York: Wiley.

Kates, R. W., and Wohlwill, J. F.
   1966   Man's Response to the Physical Environment. *Journal of Social Issues* 22, 4 (entire issue).

Kubiena, W. L.
   1962   Paleosoils as Indicators of Paleoclimates. UNESCO: *Arid Zone Research*.

LaBianca, Ø. S.
   1978   Man, Animals, and Habitat at Hesban: An integrated overview. *Andrew University Seminary Studies* 16: 229-252.

Manners, I. R.
   1969   The Development of Irrigation Agriculture in the Hashemite Kingdom of Jordan, with Particular Reference to the Jordan Valley. Unpublished doctoral dissertation, Oxford University.

Strahler, A. N.
   1973   *Introduction to Physical Geography*. New York: Wiley.

Trewartha, G. T.
   1968   *An Introduction to Climate*, 4th edition. New York: Mcgraw-Hill.

Chapter Three
BEDROCK GEOLOGY, SURFICIAL GEOLOGY, AND SOILS

Larry Lacelle

Chapter Three
Bedrock Geology, Surficial Geology, and Soils

Introduction

This chapter discusses three important aspects of the environment of the Tell Hesban area: its bedrock geology, its surficial geology (surface sediments), and its soils. The chapter is strongly oriented towards only discussing aspects of these disciplines that have, or have had, significant effects on man's use of the land. More complete, individual discipline technical presentations are referenced, should the reader wish to examine the full range of data. The history of deposition of the bedrocks and surficial materials and the development of the soils are briefly discussed in order to depict how the environment in which ancient man had to settle, came about. Emphasis is placed on the geological processes that shaped, and are still shaping, the topography of the area, and the effects of those processes on man.

Bedrock Geology

This section is limited to discussion of the bedrock geology of the Tell Hesban project area (fig. 3.1). The surficial materials and soils weathered from these bedrocks are discussed separately. The groundwater hydrology of the area is discussed in Chapter Four of this volume. This section is largely a summarization of earlier geological papers prepared for the Tell Hesban project (Bullard 1972; James 1976), but includes material from Bender's major work: "Geology of Jordan" (1974), in order to supplement the regional geologic and stratigraphic material for this section. The summarization of the above works is supplemented by the author's observations during a field expedition to the project area in 1979.

For archaeological and anthropological purposes, the well-documented geological history and characterization of the underlying bedrocks (as discussed in Bender 1974) only needs to be briefly summarized to emphasize geological data that has had an influence on man's occupation of the area. Thus, the geological characteristics and history of the area, as it affects the Tell Hesban project area, can be briefly summarized as being attributable to three primary influences (Bender 1974: 132):

1. ingressions of the ancient Tethys Sea from the west and northwest into the area of present-day Jordan. During long periods of geological history, deposits on the floor of this sea lithified (became consolidated into bedrock) into thick sequences of limestones and other related marine environment rocks. Today these rocks immediately underlie Tell Hesban (fig. 3.2);

2. the presence of the Nubo-Arabian Shield, an ancient continent from which primarily sandy sediments of continental origin were eroded, to the south. Deep strata of these sands were deposited in the area presently occupied by Tell Hesban, subsequently buried by Tethys Sea sediments, and lithified into the sandstones which today outcrop below Ain Sumiya in Wadi Hesban (figs. 3.1., 3.2); and

3. the development of the Wadi Araba Jordan River Valley zone of structural weakness. Bender suggests that this zone of subsidence has existed since Precambrian times and thus aided the ingressions of the Tethys Sea into the area. Erosion and faulting in conjunction with this subsidence has produced the steep rugged topography that characterizes the western portion of the Tell Hesban project area.

In order to relate the available geologic data to man's settlement and use of the area, the local bedrocks are discussed in order of their geographical location in relationship to the inhabitants of Tell Hesban, rather than in stratigraphic (by age) order, as would conventionally

26 ENVIRONMENTAL FOUNDATIONS

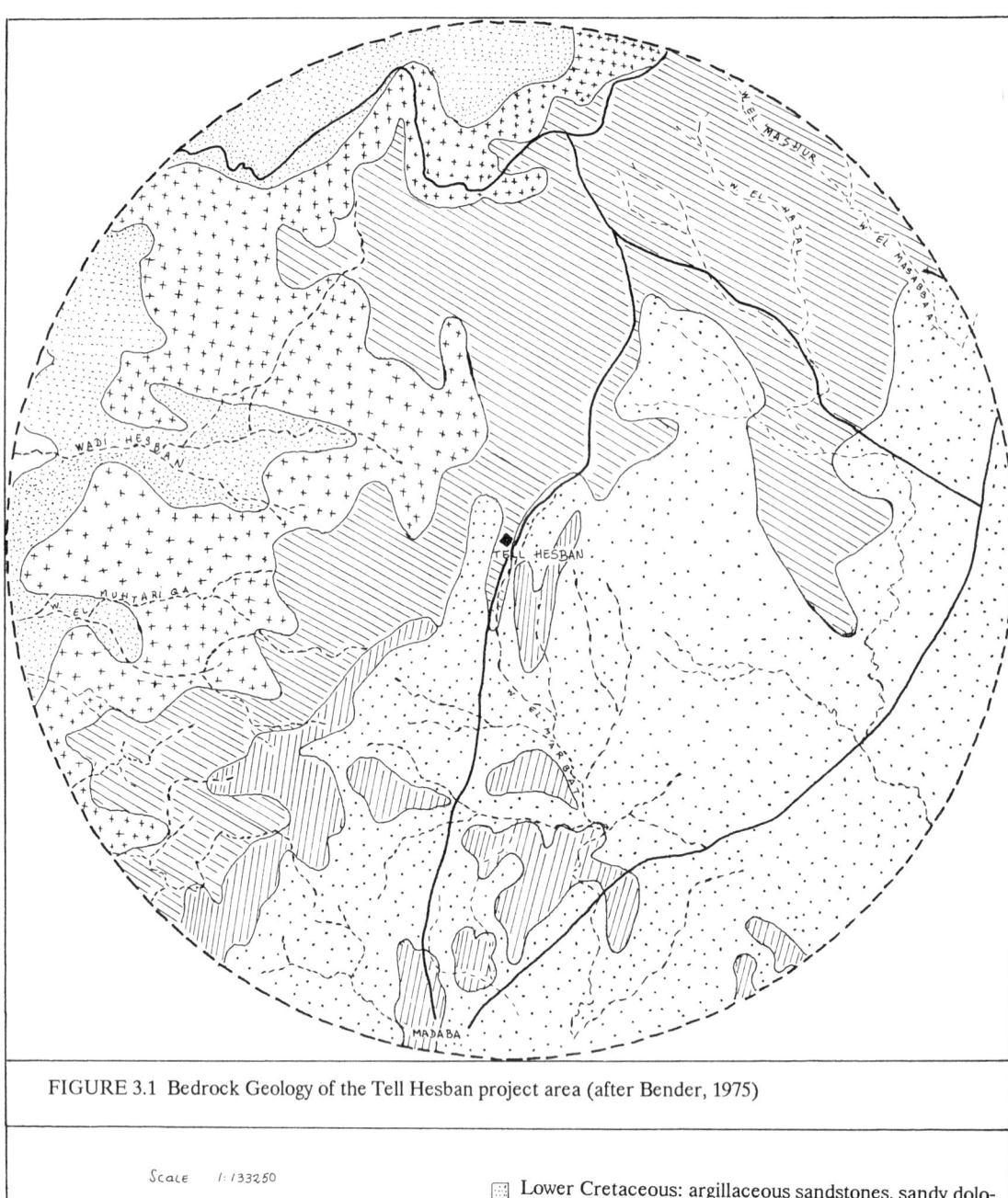

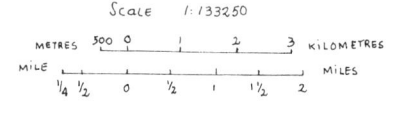

FIGURE 3.1 Bedrock Geology of the Tell Hesban project area (after Bender, 1975)

Scale 1:133250

- Lower Cretaceous: argillaceous sandstones, sandy dolomites and limestones, varicolored and white sandstones
- Upper Cretaceous: nodular and thick bedded limestones, shales and marls with gypsum, locally dolomitic
- Upper Cretaceous: chalky limestones, marls and cherts
- Upper Cretaceous: alternating limestone and coquina, locally silicified, marls and cherts
- Quaternary: terrestrial fluviatile and lacustrine unconsolidated sediments

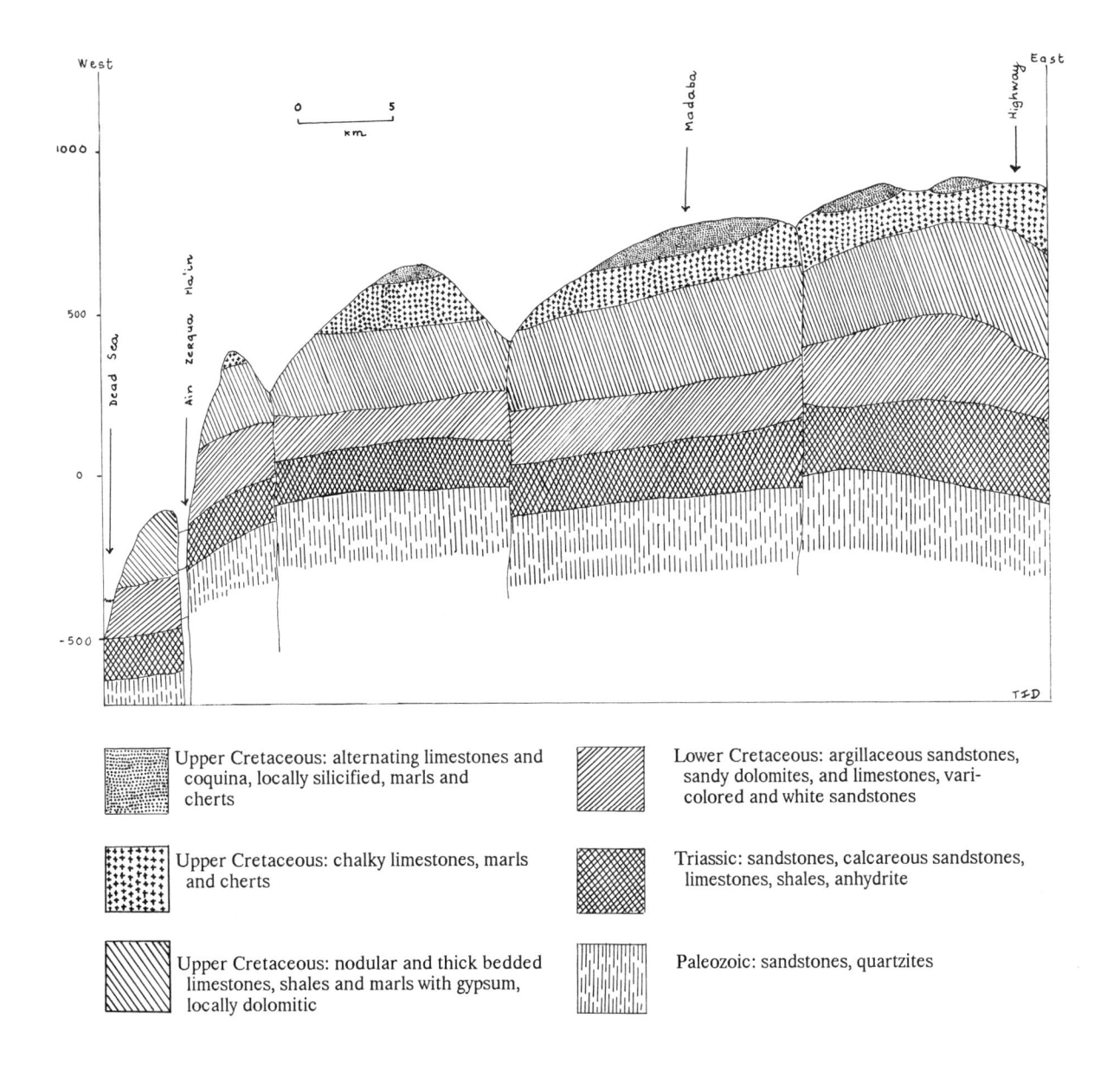

FIGURE 3.2 Geological cross-section near Tell Hesban (after Agrar und Hydrotechnik, 1977, Map HG 1.3)

Legend:
- Upper Cretaceous: alternating limestones and coquina, locally silicified, marls and cherts
- Upper Cretaceous: chalky limestones, marls and cherts
- Upper Cretaceous: nodular and thick bedded limestones, shales and marls with gypsum, locally dolomitic
- Lower Cretaceous: argillaceous sandstones, sandy dolomites, and limestones, vari-colored and white sandstones
- Triassic: sandstones, calcareous sandstones, limestones, shales, anhydrite
- Paleozoic: sandstones, quartzites

be done in a geological report. Thus, the surface *nari*, although a young deposit, is discussed first, followed by discussions of the underlying carbonate bedrocks and the sandstone bedrocks as they would be encountered in a traverse downslope from the tell. Younger (Pleistocene) conglomerates and associated weakly lithified rocks related to the infilling of the Jordan River Valley are discussed last as they are the furthest from, and likely had the least influence upon, the inhabitants of the tell.

Bedrock Characteristics

Nari

Where the carbonate bedrocks outcrop in the Tell Hesban area, their surfaces are mantled by the remains of a carbonate enriched, indurated (hardened) soil horizon that developed at the soil/bedrock interface (pl. 3.1). Erosion of the soils from the hilltops and wadi walls has exposed extensive areas of this material. The word *nari* is derived from the Arabic word *nar* (fire), as this material has been used in lime production (Horowitz 1979: 168). Elsewhere, similar materials have been referred to as caliche, calcrete, or calcareous duricrust. Where exposed, the *nari* is, in effect, the surface bedrock in the project area, as its characteristics are more rock-like, than soil-like. Thus, the physical characteristics of *nari* are discussed here, whereas its genesis is discussed in the soils section in this paper as its formation is typically part of the soil formation process in semiarid areas (Yaalon 1975: 12; Dan 1977: 78).

Nari is described by Yaalon (1975: 12-13) and Dan (1977: 69) as typically consisting of three horizons:
1. a thin, laminar crust, only 0.2 to 2 cm thick, strong (can withstand a force of 500 to 800 kg/cm^2), not breakable by hammer, nor softenable by water, but often cracked and disjointed from years of weathering;
2. an upper *nari* horizon 50 to 180 cm thick, which is relatively strong (can withstand a force of 100 to 500 kg/cm^2), and can be broken with a hammer, or scratched with a knife; and
3. A lower *nari* horizon 50 to 200 cm thick, which is softer (can withstand a force of 20 to 120 kg/cm^2), and can readily be engraved with a knife.

Bullard (1972: 134) observed that at Tell Hesban the lower *nari* horizon grades imperceptively into unweathered calcareous rocks below. Dan (1977: 79) describes weathering of *nari* exposed at the surface as generally involving runoff solution of pockets and crevices in the *nari*, often leading to formation of caves of varying shapes and sizes in the deposit. These pockets, or caves, are generally partially infilled with reddish sediments derived from the surface soils, and at occupied sites such as Tell Hesban, also generally contain grayish colored anthropogenic sediments (James 1976: 166).

Upper Mantle of Upper Cretaceous Carbonate Bedrocks

The surface of the Transjordanian Plateau near Tell Hesban is underlain by nearly flat lying, locally gently folded, or faulted, beds of resistant limestones and cherts which are interbedded with softer chalks (formed from lime mud), marls (formed from impure lime mud), and limestones (fig. 3.2) (Bullard 1972: 130). The more resistant local carbonate bedrocks are generally composed of nonporous, crystalline limestones in which much of the fossil content has been replaced by recrystallized calcite or by silica (Bullard 1972: 131) (pls. 3.2, 3.3). Such durable sediments may remain as the capping rock of a plateau remnant (pl. 3.4, arrow Z), or may stand out as a topographic shoulder on a hill (pl. 3.5, arrow) (Bullard 1972: 131). At Tell Hesban, extremely hard cherts (flintstones) are interbedded with the upper strata of carbonate bedrocks. The cherts are common as localized nodules, but at the tell also outcrop as massive beds. As they are also very resistant to weathering, their presence along with the resistant carbonate strata, likely account, in part, for the fact that the Tell Hesban hill stands out as a local topographic feature (Bullard 1972: 131). Bender's (1972: 22) description of the western edge of the Tranjordanian Plateau as being arched and dipping to the east, likely also partially accounts for the topographic high at Tell Hesban. Third, the deeply incised Wadi Majarr immediately north of the tell accentuates the tell's relief. The deepening of the Jordan River Valley over a long geological time span has resulted in concurrent deepening and headward expansion of tributary streams such as the Wadi Majarr. This wadi may have exploited a zone where the resistant capping rocks were faulted

BEDROCK GEOLOGY, SURFICIAL GEOLOGY AND SOILS 29

Plate 3.1 Nari on carbonate bedrock in the Wadi Majarr

Plate 3.2 Microphotograph of a fossililerous limestone

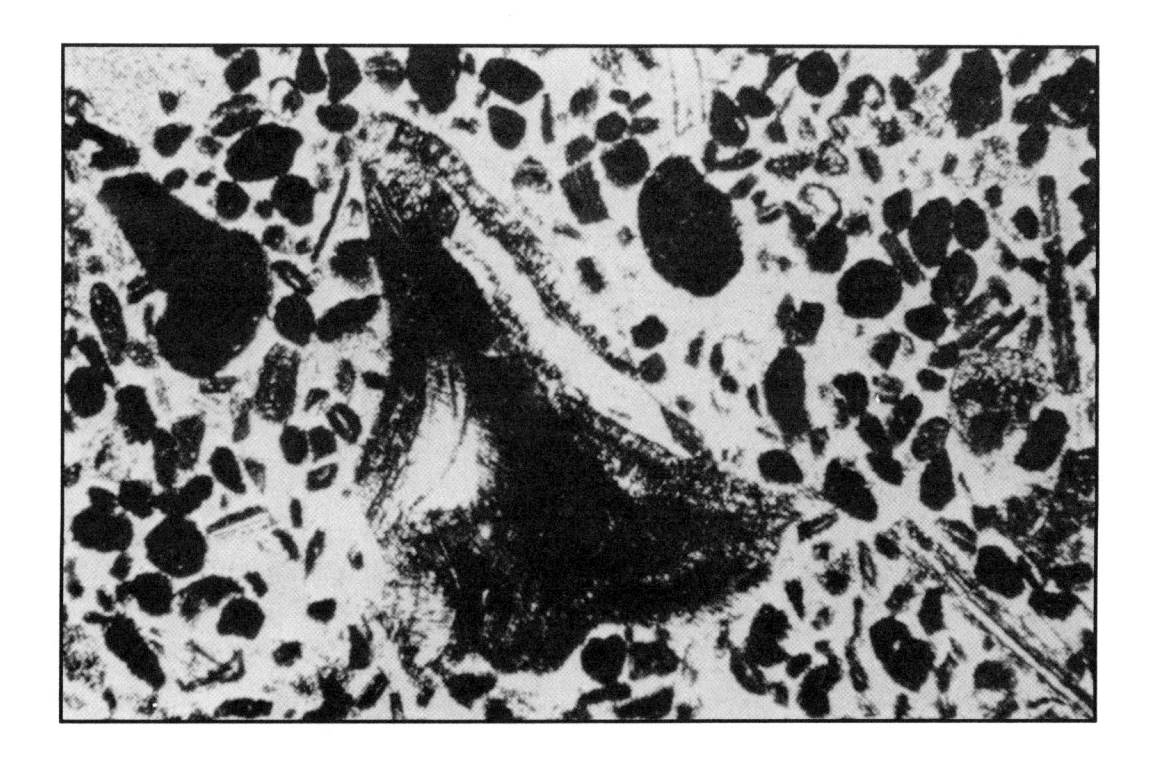

30 ENVIRONMENTAL FOUNDATIONS

Plate 3.3 Microphotograph of a very hard millstone from the surface flood on the north slope of Tell Hesban

Plate 3.4 Lithic outcrop pattern on the hill across the Wadi Majarr immediately northwest of Tell Hesban

and fractured, or it could have incised itself into an area lacking resistant strata. A fossiliferous limestone that is attractive when polished (pls. 3.6, 3.7) (classified as a pelycopodal biomicrite) outcrops at Tell Hesban (pl. 3.5, arrow). Bullard (1972: 133) also suggested that coquina (bedrock consisting almost entirely of fossils), as well as phosphate rich layers, are interbedded amongst the resistant limestones at Tell Hesban.

Lower Cretaceous and Triassic Sandstones

Friable (readily disintegrated) Lower Cretaceous sandstones are present beneath the Upper Cretaceous carbonates and are exposed in wadis Hesban and Kafrein in the project area (pl. 3.8; figs. 3.1, 3.2). Triassic (older) sandstones and related sedimentary rocks of largely continental origin occur deeper in the stratigraphic section, outcropping nearer to the floor of the Jordan River Valley.

Pleistocene Sandstones and Conglomerates

During the Pleistocene (recent) geological period, large quantities of sands and gravels were washed into the Jordan River Valley. Thus, the westernmost and lowest in elevation portions of the Tell Hesban project area (figs. 3.1, 3.2) are characterized by weakly cemented sandstones, conglomerates, and calcareous sandstones that were, in ancient times, the equivalent of the fluvial fan, terrace and floodplain deposits accumulating in the floor of the Jordan River Valley today.

Influences of Bedrock Geology on Man

Nari

Nari is the "bedrock" material that perhaps has had the greatest effect on man's use of the land in the Tell Hesban area. As the *nari* is readily workable with hand tools yet strong enough to be used as building stones, it has been extensively quarried at Tell Hesban (Bullard 1972: 135) (pl. 3.9). On the tell it was the major material used in structures and walls (Bullard 1972: 135). The relative ease of quarrying the *nari*, plus the presence of natural solution passages, resulted in extensive development of burial chambers in areas of *nari* outcrop (Bullard 1972: 134), excavation of cisterns into and through *nari* layers, plus the tooling and cleaning of natural subterranean chambers (James 1974: 169).

Upper Cretaceous Carbonates

The primarily limestone bedrocks that underlie Tell Hesban and the majority of the project area, have had a very large influence on man's use of the land. The harder local carbonate bedrocks have been quarried (pl. 3.10) and utilized in structures, walls, and road surfaces, as well as for millstones (pl. 3.11) (Bullard 1972: 136-138). Carbonate bedrocks with decorative characteristics when polished (pelycopodal biomicrite, pls. 3.6, 3.7), were locally quarried and utilized in structures, one of which was the Byzantine church on the tell (Bullard 1972: 136). Less attractive carbonate rocks, plus blocks of cherts were, and are still, utilized in structures and walls. The local abundance of cherts suitable for flaking for tool and weapon manufacture, plus springs in the western portion of the project area, must have rendered this area attractive to prehistoric man. Flint "industries" have been reported at Ain Sumiya and Mount Nebo, both within the project area (Stockton 1969).

As mentioned previously, the combination of semiarid climate and carbonate bedrocks that are subject to dissolution, absorption of surface water and subterranean passage of water, does not favor the existence of perennial surface water courses, nor waterbodies on the plateau (Chapter Four). Thus, until the advent of plaster for the sealing of cisterns, man's settlement on the plateau was severely limited by geologic factors.

The local carbonate bedrocks have also had a major influence on the physical and chemical characteristics of the soils which have weathered from them (see the soils section of this chapter). The insoluble residues left behind by the weathering of the local bedrocks have, in turn, weathered to clays with favorable water and nutrient retention properties for crop production.

The generally flat bedding of the carbonate bedrocks on the Transjordanian Plateau has also proven to be favorable for the development of agriculture, as well as for transportation. The local rocky, subdued hills that are generally attributable to geological faulting or folding, have been, and are still, ideal sites for settlement, as they were relatively easily fortified, are off the valuable agricultural lands, and have *nari* and carbonate building materials at, or near, the surface (pls. 3.4, 3.5). Short sections

32 ENVIRONMENTAL FOUNDATIONS

Plate 3.5 Tell Hesban from the northwest

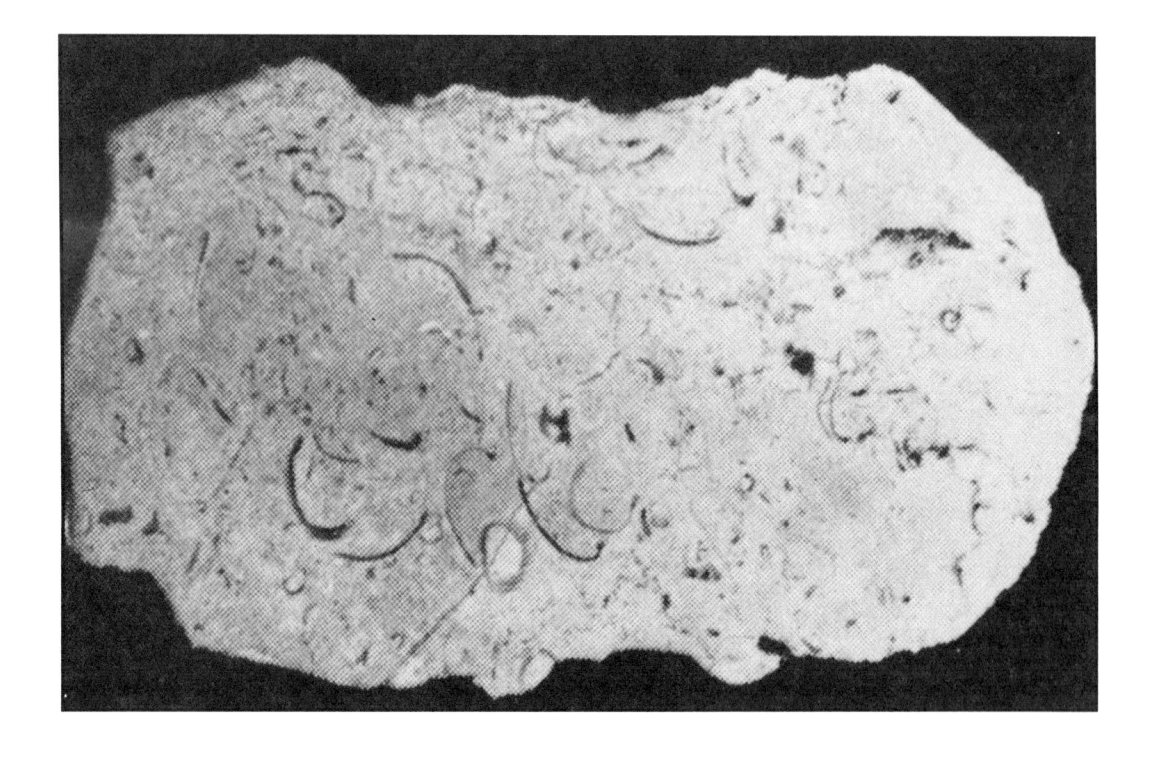

Plate 3.6 Polished section of fossiliferous limestone

Plate 3.7 Microphotograph of fossiliferous limestone

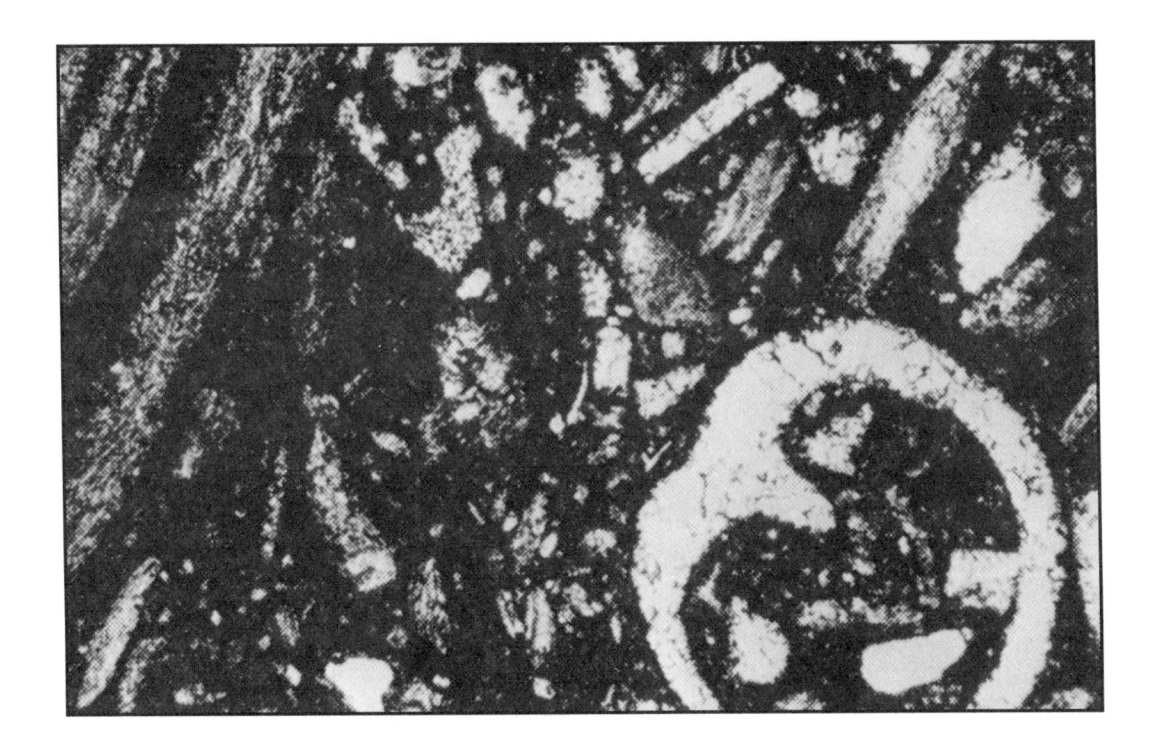

Plate 3.8 Lower Cretaceous sandstone in a cave, Wadi Nusariyat

34 ENVIRONMENTAL FOUNDATIONS

Plate 3.9 Stone block quarry on hill northwest of Tell Hesban

Plate 3.10 Partially quarried columnar section

of two ancient roads can still be observed in the project area today, despite centuries of agriculture, stone robbing and erosion. One small modern quarry suggests that limited limestone quarrying may occur locally, although quarrying activity is primarily closer to Amman. A small chalk mining operation for the production of agricultural lime was observed north of Tell Hesban at El Manshiya.

Lower Cretaceous and Triassic Sandstones

The friable sandstones that outcrop below Ain Sumiya may have been a sand source for plaster, and later, for masonry manufacture for ancient man. However, these outcrops are considerably lower in elevation than the plateau surface, and it is entirely possible that sand could have been transported to Tell Hesban on the surface of the plateau from more distant, but more accessible sources.

Pleistocene Sandstones and Conglomerates

As these weakly lithified bedrocks are on the westernmost edge of the project area and at a very low elevation, it is likely that if utilized at all, they were only utilized locally.

Pleistocene and Neogene Basalts

Although basalts do not outcrop in the Tell Hesban project area, numerous fragments of basaltic millstones, bowls, and possibly loom weights were observed in the surface detritus of the tell (Bullard 1972: 139). Thin sections taken from some basaltic artifacts (pl. 3.12) were compared to thin sections from Umm al-Jamal (pl. 3.13), a Late Roman site in an area of extensive basaltic flows near the Syrian border. The compared samples did not appear to be petrographically similar. Bullard (1972: 139-140) hypothesized that a possible source for some of the Tell Hesban basalts might have been two smaller areas of basaltic flows near the Dead Sea and much closer to Tell Hesban. However, as these flows are much lower in elevation and in rough topography, basaltic materials may have been imported from much more distant sources across the plateau surface.

Influences of Man on Bedrock Geology

Man has had considerable localized, although significant, influence in alteration of the local geology in the Tell Hesban area. His continuous landuse for settlement, grain growing and grazing has, in all likelihood, accelerated natural rates of erosion and contributed to the exposure of extensive areas of *nari* and bedrock that might otherwise still be protected by a mantle of soil. Man has altered the surface shape of Tell Hesban by quarrying (pl. 3.4) (Bullard 1972: 136). Natural caves and conduits in the *nari* and carbonate bedrock have been modified by man for uses including: water storage, burial, animal shelters, and probably habitational and food storage use. Natural drainage passageways have been altered and new routes constructed as a result of man's demand for water storage, sewage drainage, refuse dumping, wall building, and street construction (James 1974: 165).

Whereas preoccupational cave sediment resembles the surface soil in color, postoccupational cave sediments are generally much more unevenly sorted, and are gray in color. The discoloration of this sediment is attributed to man's domestic activities, including the incorporation of charcoal into the soil (James 1974: 166).

The influences of bedrock geology on man, and man on bedrock geology, are further discussed in Chapter Eight of this volume.

Surficial Geology

Surficial geology, as discussed in this section, involves the description of the sediments overlying the bedrock at the surface of the Earth. It includes description of their physical characteristics, characteristic landforms, processes of deposition, and influence on man's use of the land. The bedrock that they have weathered from and the soils weathered from them are discussed elsewhere in this chapter. The surficial geology of the Tell Hesban area has also been discussed in Bullard (1972), and James (1976).

Physiographic Subdivisions and Landforms

Bender (1974) describes three physiographic subdivisions in the Tell Hesban project area: (1) the East Jordanian limestone plateau (the plateau); (2) the highlands at the eastern rim of the Wadi Araba-Dead Sea-Jordan River Grabben (the hills); and (3) the Wadi Araba-Dead Sea-Jordan River Depression (the wadis) (fig. 3.3).

36 ENVIRONMENTAL FOUNDATIONS

Plate 3.11 Grinding wheel of hard fossiliferous limestone

Plate 3.12 Microphotograph of a fragment of a basalt bowl

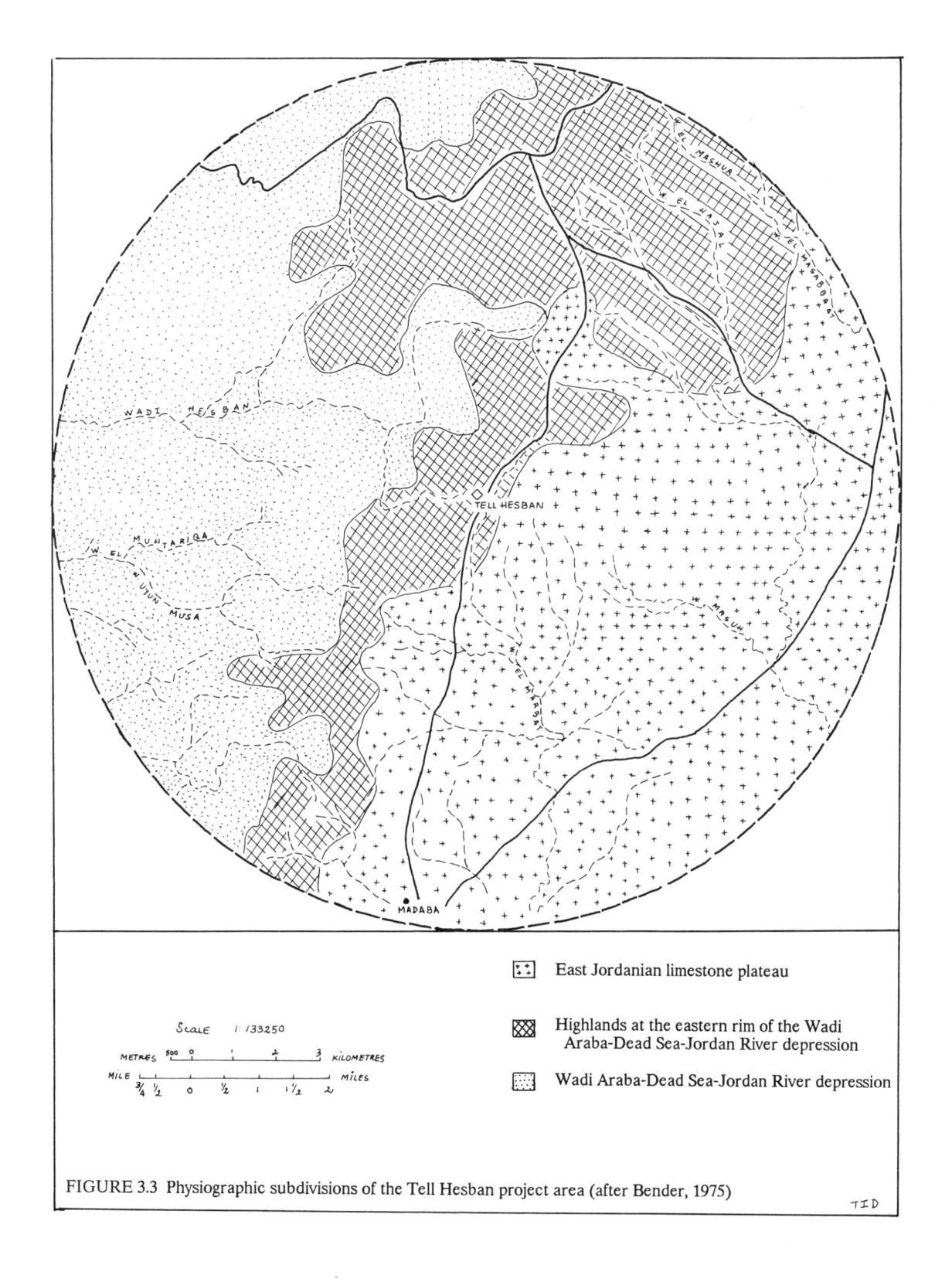

BEDROCK GEOLOGY, SURFICIAL GEOLOGY AND SOILS 37

FIGURE 3.3 Physiographic subdivisions of the Tell Hesban project area (after Bender, 1975)

The plateau surface is characterized by gentle relief with extensive plains interrupted locally by low bedrock topped hills (pl. 3.14). Material eroded from the hill summits has accumulated in the troughs between hills, forming gently sloping, relatively stone-free plains (fig. 3.4). The gradients of the intermittent streams on the plateau surface are generally low and the channels are not deeply incised. Neither the hills nor the plains show evidence of gullying, although closer to the edge of the Jordan River Valley the gully heads of the steep wadis reach onto the plateau.

The hills at the edge of the Jordan River Valley are not extensive nor high at Tell Hesban, but are more evident at Naur at the northern edge of the project area. Here, the hills are more numerous, slopes steeper, and a much greater proportion of bedrock is exposed. The steep walled wadis leading down to the Jordan River partially dissect this hilly area.

The slopes leading down to the Jordan River have been severely dissected by wadi erosion, resulting in the development of numerous, deep, V-shaped, trellised canyons (pl. 3.15). Ridges between wadis are commonly steep sided with sharp summits. However, in many areas, isolated plateau remnants or round topped hills separate the wadi networks (fig. 3.5).

The steep slopes leading down to the floors of the wadis, and much of the hill and plateau tops, consist primarily of exposed bedrock. Shallow, locally weathered surficial materials veneer portions of the slopes and fill pockets in the bedrock. Fluvial fan, fluvial terrace, and, to a lesser degree, colluvial veneers are evident on the lower slopes of the wadis (pl. 3.16; fig 3.6).

Geomorphic Processes and Physical
Characteristics of the Surficial Materials

On the surface of the plateau, solution weathering of the carbonate bedrocks has left residues of finely textured sediments along with lesser amounts of insoluble cherts. Many years of physical weathering coupled with soil forming processes have resulted in the finer particles weathering to kaolinite and illite clays (Bullard 1972: 132). Not being in an area characterized by extensive erosion or sedimentation, it is likely that these clay rich sediments are very old. Several investigators suggest that the soils developed on these sediments are relic soils from Pliestocene pluvial periods (Moorman 1959; Butzer 1961: 449; Horowitz 1979: 167-168). If the soils on these materials date back to the ice ages, then it is probable that the surficial materials are older.

The denudation of the tops of the hills in this area was attributed by Thompson (1880: 642) and Moorman (1959) to the soil being washed off the hillsides. However, field examination revealed that evidence of surface erosion processes such as gullying is uncommon on the surface of the plateau, except where vigorously headward cutting wadis of the Jordan River Valley have extended inland onto the plateau. Amiran and Gilead (1954: 286) suggest that heavy textured soils such as *Terra Rossas* on limestones (as at Tell Hesban) are the least susceptible Palestinian soils to gullying. Atkinson *et al.* (1967) further suggest that the relatively low precipitation on the plateau (150 to 400 mm per year in the project area), coupled with the measured ability of the soil to infiltrate amounts of precipitation greater than maximum observed rainfall intensities, also aids in prevention of surface erosion and gullying. Fleming and Johnson (1975) observed soil creep averaging 10 mm per year under somewhat similar conditions in California (approximately 400 mm precipitation per year, grass and scattered oak vegetation, montmorillonite rich (swelling) clays and 12 to 14% slopes). Although Tell Hesban surficial materials are not as rich in montmorillonite, they are likely susceptible to a similar creep process, but likely at a lesser rate. However, when the probable age of the sediments at Tell Hesban is considered, such a process appears to be sufficient to move materials downslope off the hilltops (pl. 3.17).

Intermittent stream courses, often incised less than 1 m into the swales between hills, are commonly bedded with course fragments on their flat beds. In addition to the reasons discussed above, the lack of more extensive, or deeper, gullying on the plateau may be partially attributable to the lack of large catchment basins to funnel flow into the watercourses.

The highlands along the edge of the plateau show greater evidence of slope denudation than inland, as the topography here is steeper, wadi incision common, and precipitation is maximum for the project area (pl. 3.18). Denudation of the hills here has likely been aided by removal of material from the lower slopes by rapid wadi

Plate 3.13 Microphotograph of a fragment of a basalt bowl

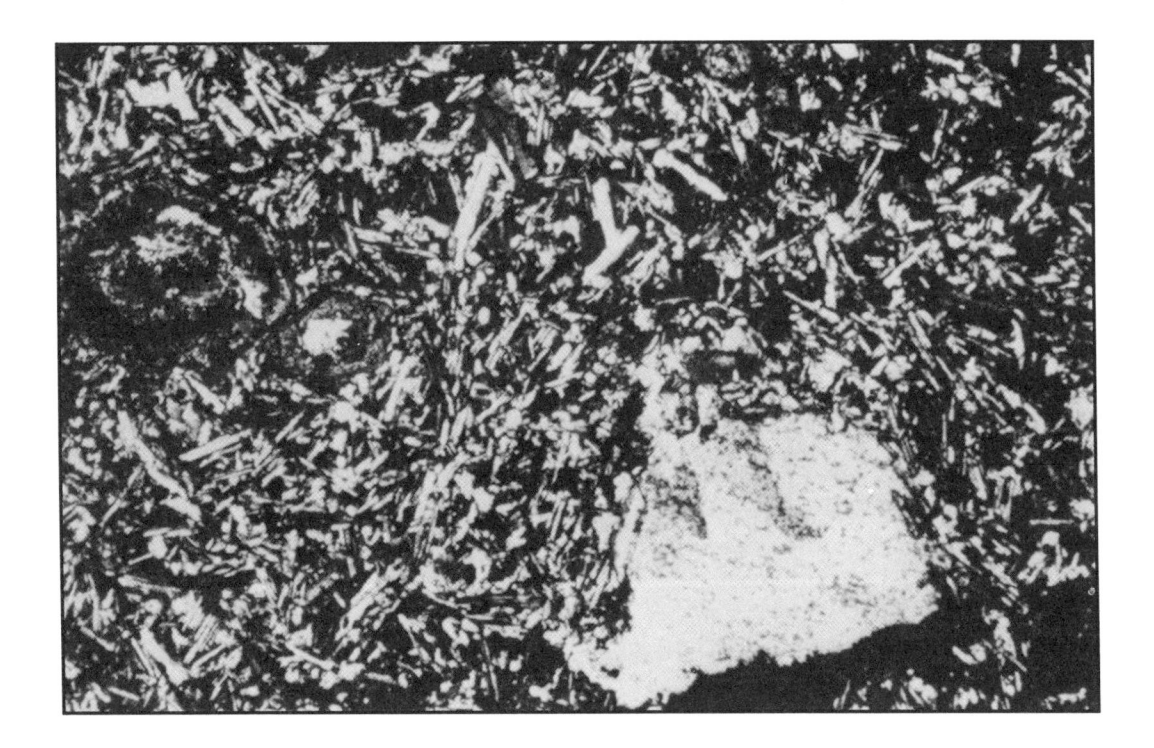

Plate 3.14 Topography of the Transjordanian Plateau

FIGURE 3.4 Cross Section depicting surficial materials on the surface of the Transjordanian Plateau

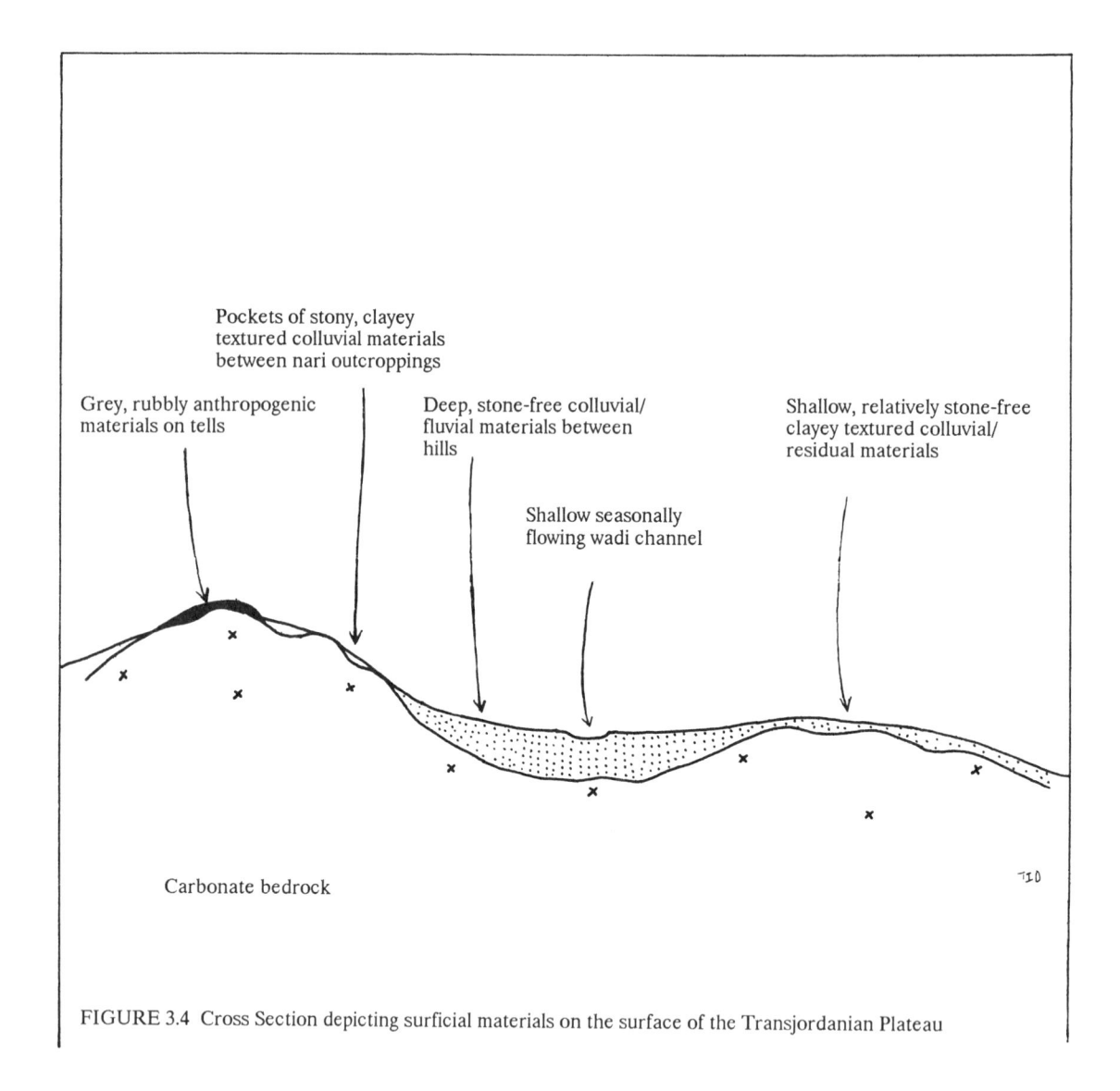

BEDROCK GEOLOGY, SURFICIAL GEOLOGY AND SOILS 41

Plate 3.15 Deep, steep walled wadis west of Tell Hesban

Plate 3.16 Fluvial sediments in a wadi below Tell Hesban

42 ENVIRONMENTAL FOUNDATIONS

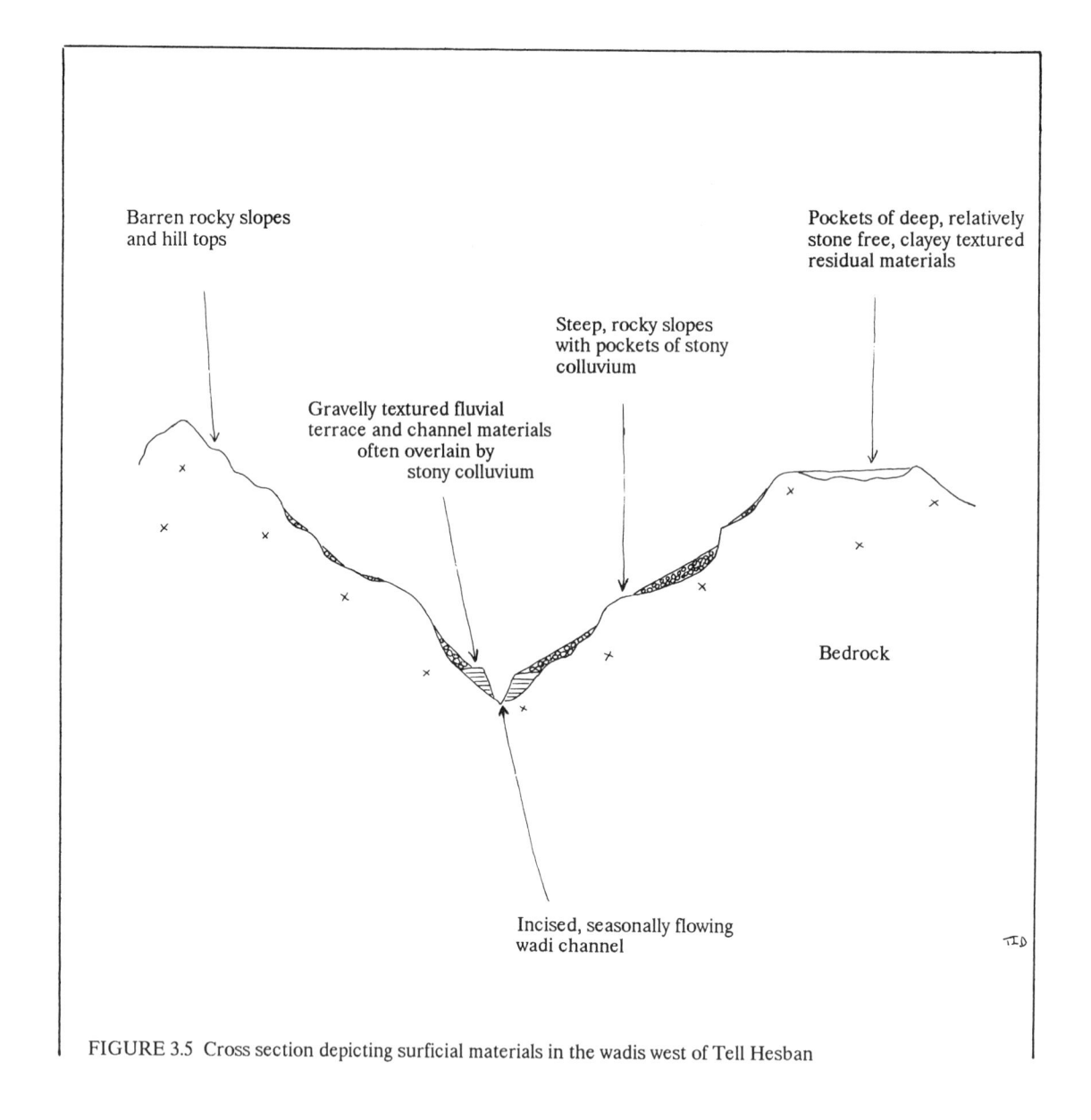

FIGURE 3.5 Cross section depicting surficial materials in the wadis west of Tell Hesban

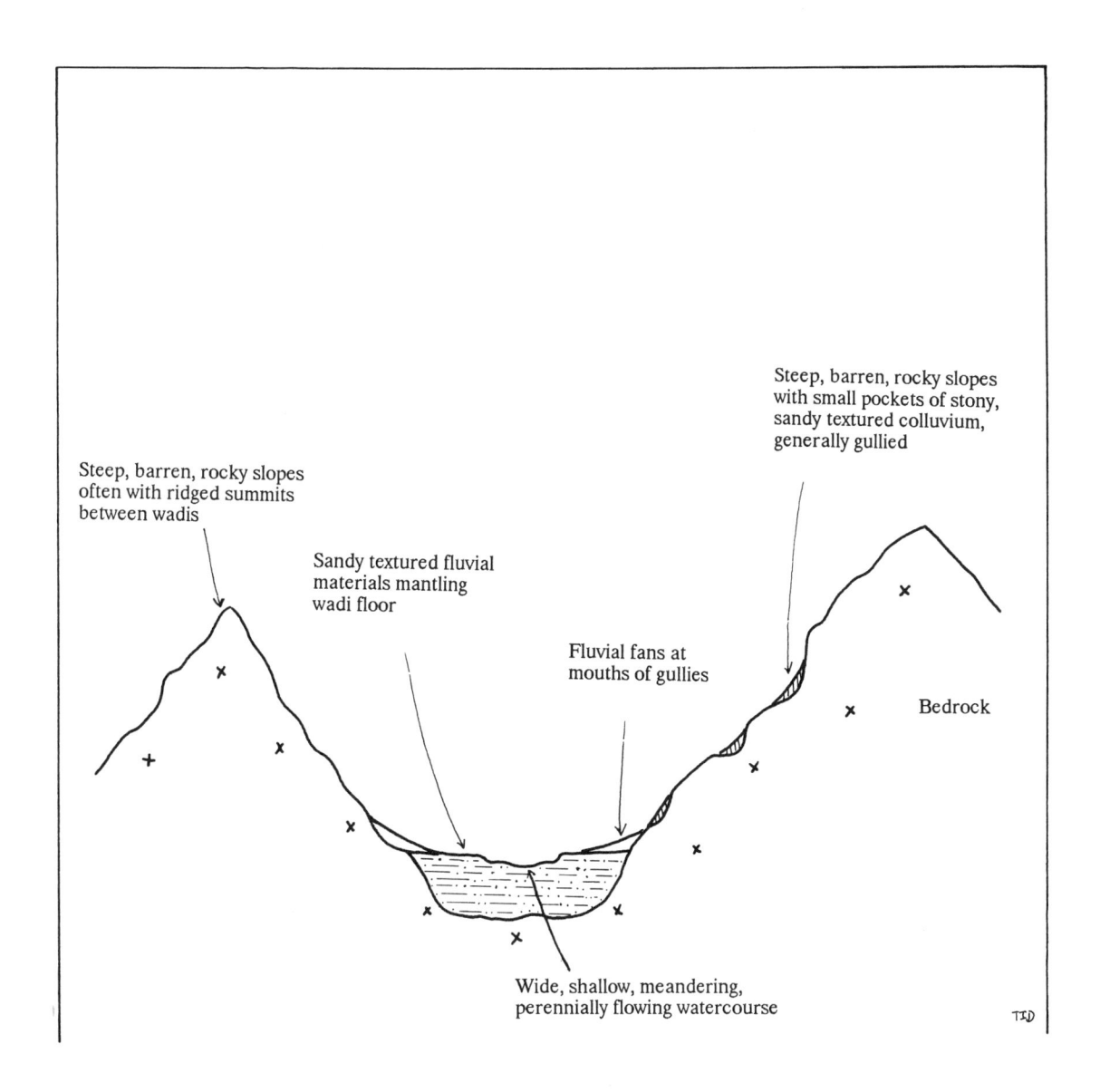

FIGURE 3.6 Cross section depicting surficial materials in the wadis near the floor of the Jordan River Valley west of Tell Hesban

44 ENVIRONMENTAL FOUNDATIONS

Plate 3.17 Colluvium mantling lower slopes of a hill near Tell Hesban

Plate 3.18 Barren hillsides in the wadis west of Tell Hesban

growth, especially when occasional heavy rainfalls cause flash flooding. In these highlands, extensive deposits of residually weathered clay rich surficial materials are less common than on the plateau. In many areas, the clays are confined to pockets in the bedrock. Coarser textured, shallow surficial materials, more recently weathered from the local bedrocks, are common on the slopes of the highlands. These materials often contain a relatively high proportion of course fragments *(nari,* limestone, and chert).

Fluvial erosion is the dominant geologic process in the wadis leading down to the Jordan River. Stream beds locally exhibit erosive scouring into the bedrock. Tumble-polished boulders of over one m maximum diameter were observed as part of the streams bedload (Bullard 1972: 131). Localized pockets of generally shallow, compact, clay rich sediments may be found on the hills and plateau remnants in this area. However, the majority of the areas that are not exposed bedrock are veneered by loose, coarse textured, locally weathered surficial materials containing numerous bedrock fragments. The floors of wadis often contain narrow strips of terrace or fan remnants. These are crudely stratified and contain high percentages of sands and coarse fragments. Where the wadis have dropped down to the elevation of the sandstone bedrock outcroppings, the surficial materials become proportionately more sandy.

Influences of Surficial Geology on Man

Bullard (1972: 132) reported that the local clays, being dominated by kaolinite with lesser illite, offer ". . . excellent potential for ceramic clay sources. . . ." He also observed that the terrace deposits in the local wadis are highly suited for mud brick manufacture. Lack of deposits of clean sands on the plateau surface undoubtably had a significant impact on the settlers in the area during the historical periods when plaster and later, mortar, were used.

Influences of Man on the Surficial Geology

The previous discussion of hill denudation partially supplies an answer to the question often asked—whether the present-day partially denuded landscape in the Tell Hesban area is attributable to man's activities. It appears that natural geological processes could result in the degree of erosion that is today evident. However, as land degradation due to farming and grazing practices can readily be seen throughout the world, it is probable that these activities contributed significantly to the current state of denudation evident in the Tell Hesban area today.

Soils

This section describes the soils of the Tell Ḥesbân project area (fig. 3.7) in terms of: (1) their classification and genesis; (2) their physical and chemical properties; and (3) their influence on man. The bedrock and surficial materials from which the soils have been derived have been previously discussed. Soil surveys covering the Tell Hesban project area, or areas adjacent to it, include Moorman (1959), Atkinson *et al.* (1967), and West (1970). As only limited field data was collected by the author, most of the data reported in this section is a summarization from the above reports.

Soil Classification

Soils in the Tell Hesban area are described in this section according to their technical classification, using the system employed by Moorman (1959) in his reconnaissance survey of Jordan. The history of development of the local soils, the physical and chemical properties of the soils, and the influence of the soils on man are then discussed in terms of Moorman's major classes of soils. His system of classification is based on the 1938 United States Department of Agriculture system. Although not the most current, Moorman's system has been used in this section as it is descriptive and often referred to in the available literature for the area. Equivalent nomenclature according to the more modern American classification (Soil Survey Staff 1975) and the Food and Agriculture Organization of the United Nations Soil Map of the World (1974) is given for comparison.

Moorman (1959) refers to the soils of the highlands at the western edge of the Transjordanian Plateau and adjacent plateau as Red Mediterranean Soils (figs. 3.7, 3.8; pl. 3.19). He differentiates these soils from the very similar *Terra Rosa* Soils commonly mapped in more humid Mediterranean areas, on the basis of the Red Mediterranean Soils commonly being calcar-

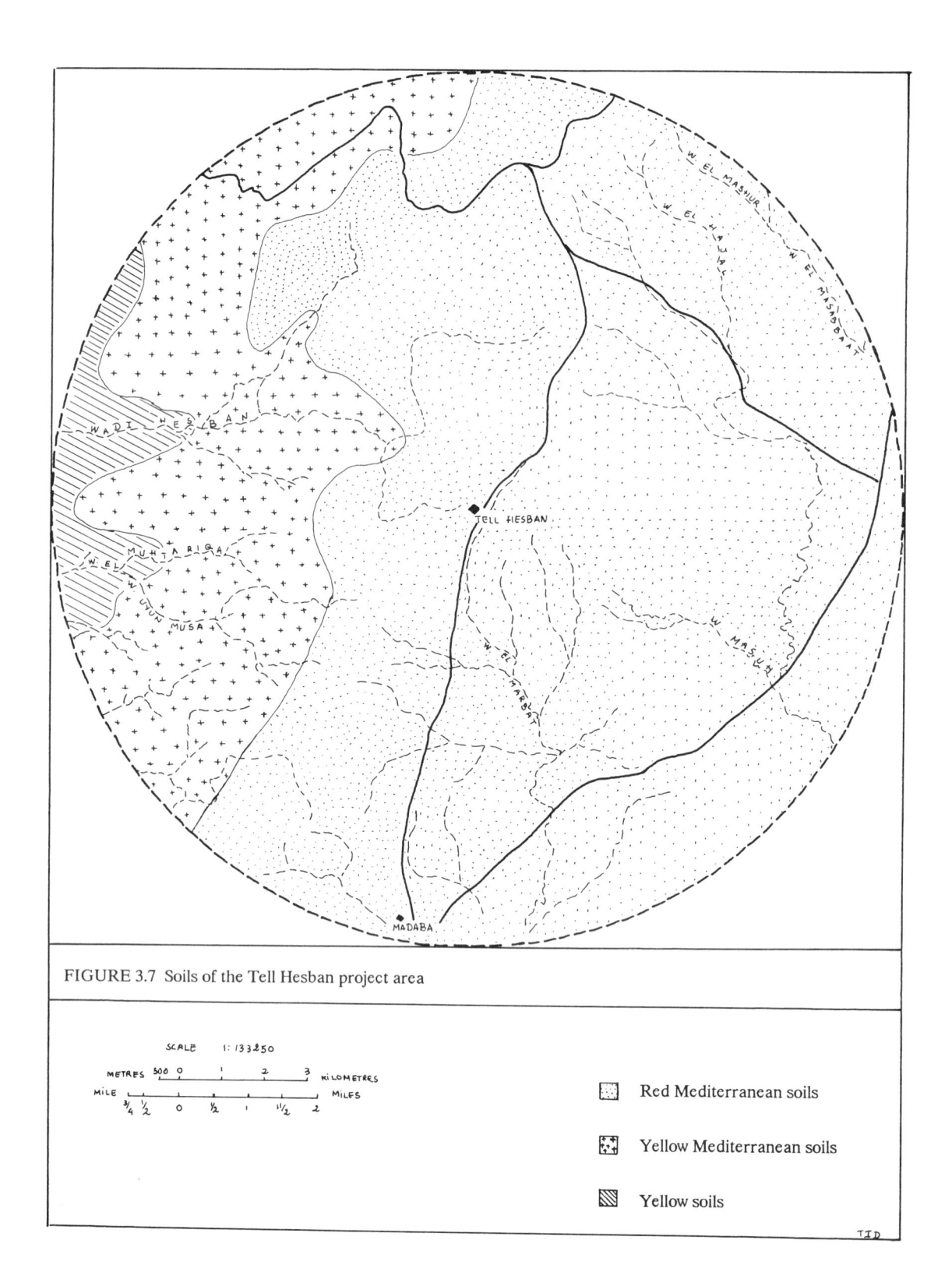

FIGURE 3.7 Soils of the Tell Hesban project area

SCALE 1:133,250

- Red Mediterranean soils
- Yellow Mediterranean soils
- Yellow soils

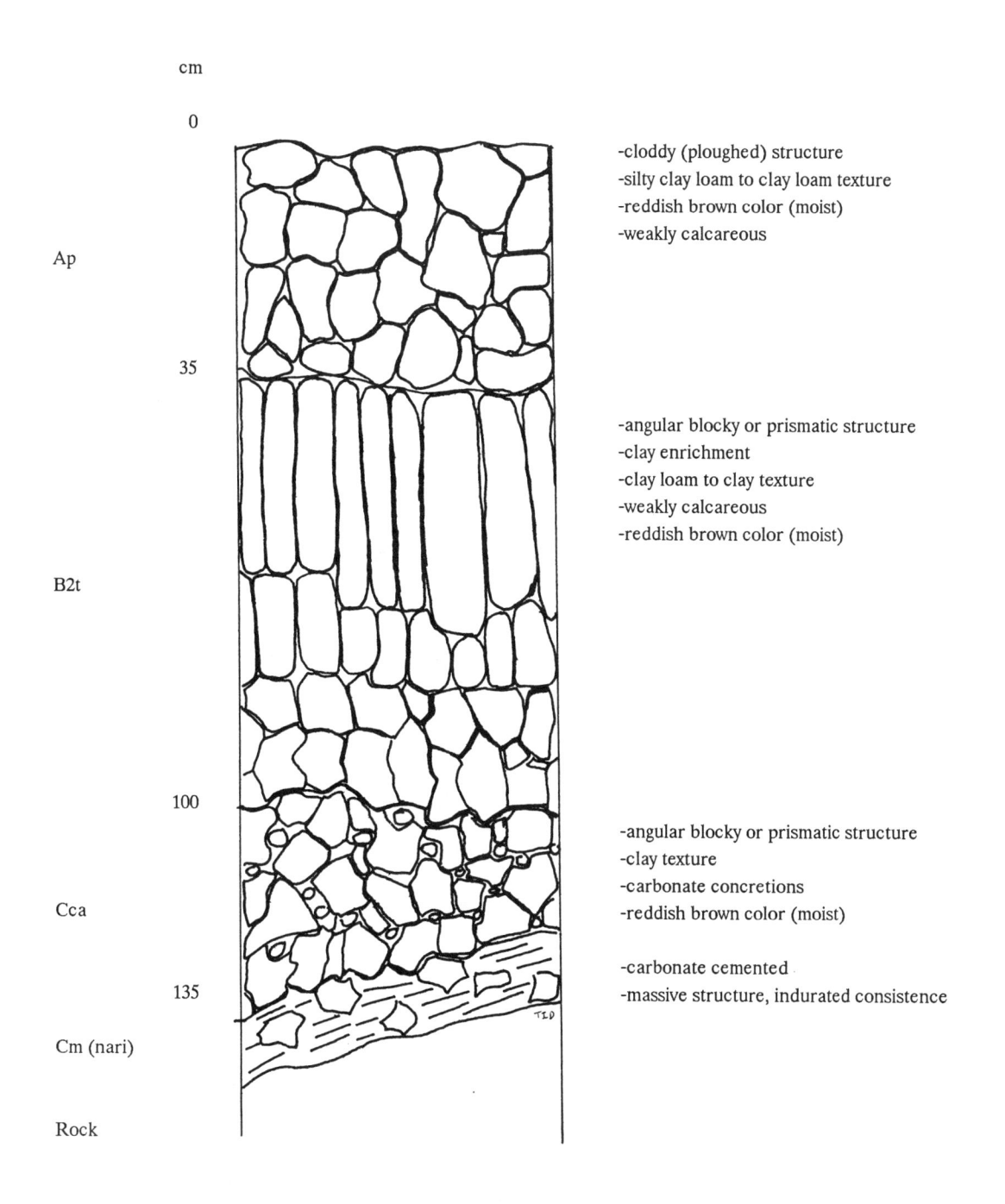

Horizon	Depth (cm)	Description
Ap	0–35	-cloddy (ploughed) structure -silty clay loam to clay loam texture -reddish brown color (moist) -weakly calcareous
B2t	35–100	-angular blocky or prismatic structure -clay enrichment -clay loam to clay texture -weakly calcareous -reddish brown color (moist)
Cca	100–135	-angular blocky or prismatic structure -clay texture -carbonate concretions -reddish brown color (moist)
Cm (nari)	135+	-carbonate cemented -massive structure, indurated consistence
Rock		

FIGURE 3.8 Profile of a typical Red Mediterranean Soil

eous to the surface. Red Mediterranean Soils may also be classified as Calcic Rhodoxeralfs (Soil Survey Staff 1975) or Chromic Luvisols (Food and Agriculture Organization of the United States 1974). Moorman describes the less brightly colored, although in most other ways similar, soils of the drier areas in the wadis leading down to the Jordan River and in the drier areas further east on the plateau, as Yellow Mediterranean Soils (figs. 3.7, 3.9; pl. 3.20). The Yellow Mediterranean Soils appear to have similar equivalent classifications in the 1975 American and 1974 United Nations systems, to those of the Red Mediterranean Soils. For the portion of the Tell Hesban project area that is on the lower slopes of the Jordan River Valley, Moorman (1959) describes weakly developed Yellow Soils (figs. 3.7, 3.10). Yellow Soils are characteristic soils of the eroded bedrocks of that area. In the 1975 American system, the Yellow Soils appear to correspond to Typic Camborthids and, in the 1974 United Nations system, to Calcic Yermosols.

Moorman also describes two other azonal (occurring in all zones) soils: (1) Alluvial Soils on the floodplains and fans of the wadis (pl. 3.21); and (2) Regosolic Soils (young soils lacking distinct soil horizons attributable to soil development), found primarily on areas of exposed *nari* that have recently weathered to produce coarse textured soils. In this report, tell soils have been distinguished as a type of Regosolic Soil of special interest to archaeologists (pl. 3.22).

Soil Genesis

The processes of formation of the local soils are discussed in this section in order to clarify how the local soils acquired their distinctive physical and chemical properties, and to shed further light on the long term environmental history of the area. Horowitz (1979: 167) suggests that the *Terra Rossas* of the mountains in Israel (and therefore likely also the very similar soils at Tell Hesban) are paleosoils (relic soils, reflecting an ancient soil forming environment). He suggests that the noncalcareous A (surface) soil horizons of typical *Terra Rossas* have been removed by erosion (in the Tell Hesban area, thus creating conditions for Moorman's (1959) Red Mediterranean Soil). Moorman (1959), Butzer (1961: 45), and Horowitz (1979: 167) all cite evidence to support the argument that such soils developed during more humid periods in the Pleistocene.

There are three dominant processes that act in the formation of *Terra Rosa* type of soils. They are: (1) the precipitation of calcium carbonate (commonly referred to as lime) in the subsoil; (2) the accumulation of clays in the B (mineral soil) horizon; and (3) the development of iron staining in the B and A soil horizons (fig. 3.8).

Yaalon (1975: 12-13), and Dan (1977: 77-79) describe the origins of the *nari* (discussed in the bedrock geology section of this chapter) as being of pedogenic (soil forming processes) origins. Dan describes the process as beginning with limited lime solution and its redeposition in the transitional layer between the soil and the bedrock (fig. 3.8). This secondary lime accumulation produces a indurated (hardened) layer on the rock surface. When the soil has been eroded away, this layer becomes the *nari* so evident in the Tell Hesban area today. Dan (1977: 79), and Horowitz (1979: 168) cite evidence to suggest that the *nari* seen today originated as a soil horizon in Pleistocene times, or in some cases, earlier. Their evidence further substantiates an ancient age for the Red and Yellow Mediterranean soils of the Tell Hesban. The genesis of *Terra Rosa* type of soils is the accumulation of clays in the B horizons (fig. 3.8). The clays coat solid peds (aggregates of soil particles) resulting in eventual formation of one or more of: columns, multi-sided prisms, angular or subangular blocks. The third significant soil forming process involves disintegration of iron bearing heavy minerals into reddish colored sesquioxides (Horowitz 1979: 167), thus giving these soils their characteristic color.

The soil forming processes of the Yellow Mediterranean Soils are apparently similar, but generally appear to be less fully expressed (fig. 3.9). The drier environment that these soils developed in apparently was not as favorable for lime, clay, and iron alterations. The Yellow Soils adjacent to the floor of the Jordan River Valley are commonly weakly developed soils on recently exposed bedrock, or on recently deposited colluvial or fluvial deposits. Even when the soil parent materials are much older in this area, soil developments tend to be weak, as soil forming processes are not very active in such arid environments. Thus, the carbonate enriched layer(s) on Yellow Soils are generally closer to

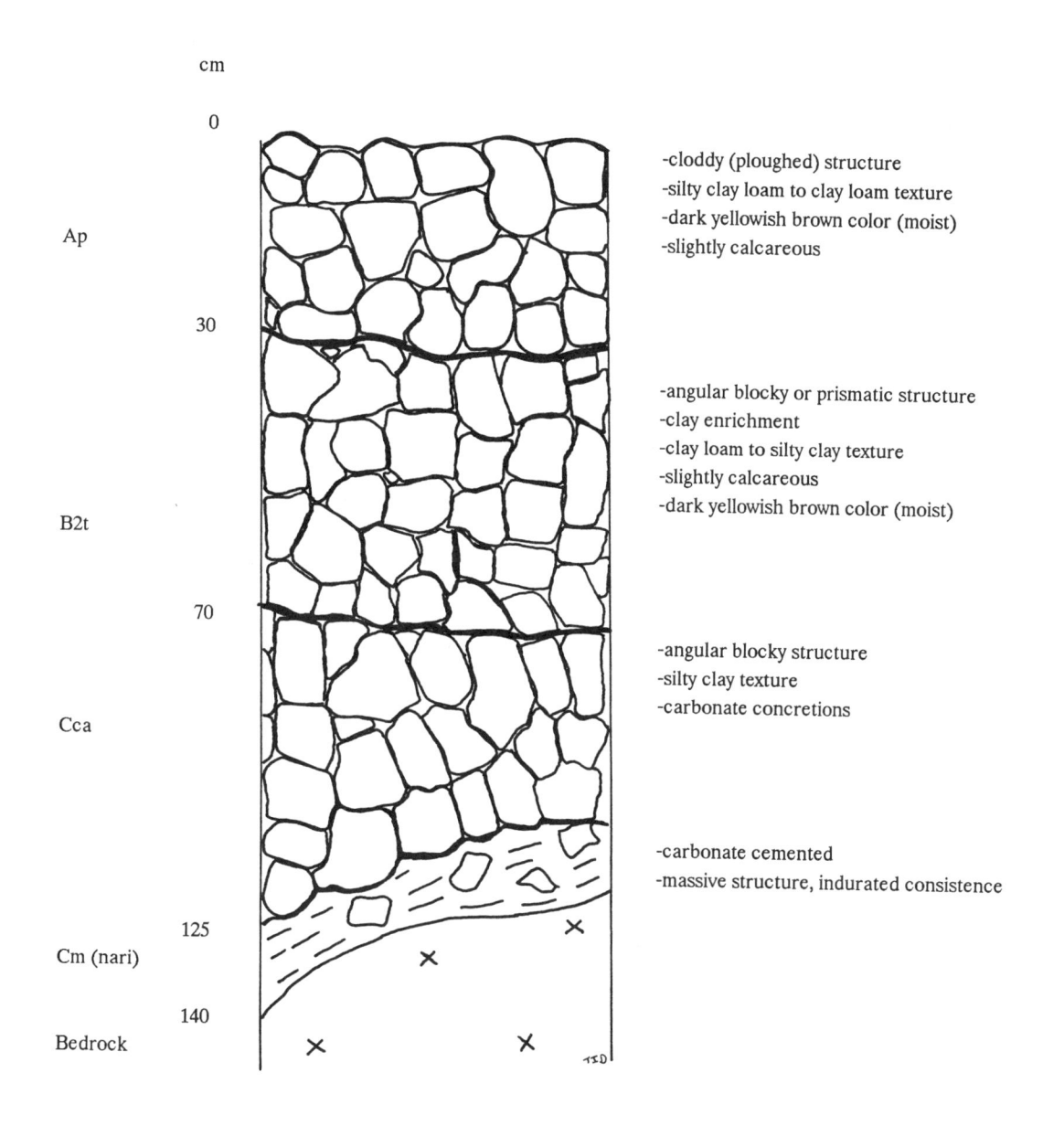

```
                cm
                 0
                                    -cloddy (ploughed) structure
                                    -silty clay loam to clay loam texture
     Ap                             -dark yellowish brown color (moist)
                                    -slightly calcareous

                30

                                    -angular blocky or prismatic structure
                                    -clay enrichment
                                    -clay loam to silty clay texture
                                    -slightly calcareous
     B2t                            -dark yellowish brown color (moist)

                70
                                    -angular blocky structure
                                    -silty clay texture
                                    -carbonate concretions
     Cca


                                    -carbonate cemented
                                    -massive structure, indurated consistence
                125
     Cm (nari)
                140
     Bedrock
```

FIGURE 3.9 Profile of a typical Yellow Mediterranean Soil

50 ENVIRONMENTAL FOUNDATIONS

Plate 3.19 Red Mediterranean Soils on the surface of the Transjordanian Plateau

Plate 3.20 Yellow Mediterranean Soils on hilly topography

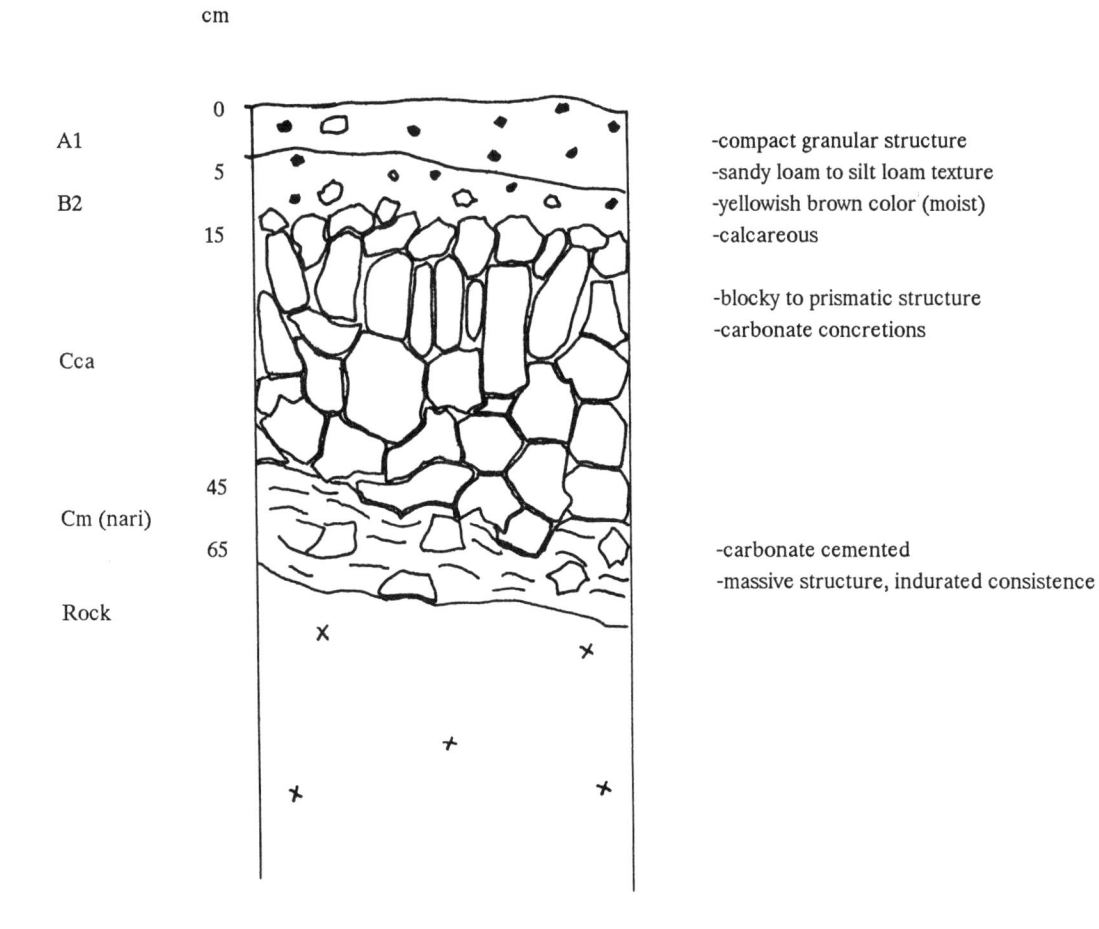

Depth		
A1	0	-compact granular structure
	5	-sandy loam to silt loam texture
B2		-yellowish brown color (moist)
	15	-calcareous
		-blocky to prismatic structure
		-carbonate concretions
Cca		
	45	
Cm (nari)		
	65	-carbonate cemented
Rock		-massive structure, indurated consistency

FIGURE 3.10 Profile of a typical Yellow Soil

52 ENVIRONMENTAL FOUNDATIONS

Plate 3.21 Shallow young soils on *nari*

Plate 3.22 Profile of Tell Hesban soils

the surface due to less deep soil weathering. Regosolic Soils on the erosion denuded hills of the plateau and highlands, as well as on fluvial deposits in the wadis, lack well-developed soil horizons due to their youth.

Soil Physical Properties

Physical soil data discussed in this section are based upon the field observations of the author, supplemented by data from Moorman (1959), Atkinson *et al.* (1967), and West (1970).

The dominant physical characteristic of the Red Mediterranean Soils is their high clay contents (50 to 70% [Atkinson *et al.* 1967], 50 to 80% [West 1970]) (table 3.1). It is this high clay content which accounts for the considerable ability of these soils to infiltrate water (14 mm per hour [Atkinson *et al.* 1967]), and retain it for plant use (Moorman 1959; Atkinson *et al.* 1967). Their indurated carbonate rich subsoil horizons are also significant as they are largely impermeable and thus limiting on the depth of soil available for water storage. The Red Mediterranean Soils are plastic when wet and consequently sticky and heavy to work with. When dry, the B and C horizons solidify into distinct, hard, subangular blocky, angular blocky, prismatic or columnar soil structure (fig. 3.8). On barren ground and on plowed fields the soil surface commonly cracks open when the soil becomes desiccated, a process which damages plants, but aids in quick infiltration of large quantities of rainfall during heavy storms.

The Yellow Mediterranean Soils generally have similar physical properties as the Red Mediterranean Soils, but are characteristically yellow-brown rather than reddish-brown in color. Both the often relatively sandy-textured Yellow Soils, and the shallow Regosoic Soils with their relatively high percentages of coarse fragments (due to having recently weathered from *nari* and impure limestones), have much lower abilities to retain soil water than do typical Red or Yellow Mediterranean soils. They also typically have a loose soil consistence, weak soil structure, and are nonplastic, thus making them much easier to work whether wet or dry (table 3.1).

Soils developed on fluvial deposits in the wadis are commonly characterized by sand and gravel textures, single grain structure, and lack of compaction. They are very permeable and have little ability to retain water for plant nutrition. Soils of the tells and villages are generally characterized by gray, relatively sandy soil textures, high percentages of angular coarse fragments (archaeological rubble), and variable degrees of compactness.

Soil Chemical Properties

As no chemical soil analyses were done for the Tell Hesban project, the data reported in this section are summarized from reports by Moorman (1959), Atkinson *et al.* (1967), and West (1970).

The most important chemical soil property of Red Mediterranean Soils likely is their relatively high level of bases, especially potassium and calcium (both important nutrients for plant growth) (table 3.2). They are commonly calcareous to the surface and consequently have a high pH (neutral to basic). They are not usually high in salts, but are deficient in organic matter.

The Yellow Mediterranean Soils are generally similar in chemical composition to the Red Mediterranean Soils, except for commonly being more calcareous (Moorman 1959). The Yellow Soils, being in a topographically lower position and therefore subject to downslope seepage, commonly have high lime and salt contents. Consequently, they are generally low in the other bases needed for plant nutrition.

Soils developed in surficial materials weathered from sandstone bedrocks may be noncalcareous and nonsaline, but in most areas mixing of materials and downslope seepage results in their becoming calcareous. In addition, they commonly have low nutrient and organic matter contents.

Alluvial Soils developed on the floodplains, fluvial fans, and terraces in the wadis are generally strongly calcareous. They are also deficient in nutrient status and organic matter content.

Regosolic Soils on recently deposited colluvium, or on materials recently weathered from bedrock, are commonly strongly calcareous, and being relatively coarse textured, are also low in organic matter and nutrient status.

Tell soils and modern village soils are commonly strongly calcareous and high in nutrient status due to refuse incorporation.

Table 3.1

TABLE 3.1 Soil physical properties (after Atkinson 1967 and the author's observations)

Soil Development	Horizon	Parent Material	Texture	Stoniness	Structure	Permeability	Plasticity	Color
Red Mediterranean Soil	A B C	residual	clay loam clay ----	non-stony non-stony variable	prismatic or blocky blocky	slow slow slow	plastic plastic plastic	reddish brown ---- ----
Yellow Mediterranean Soil	A B	colluvium	---- silty clay	non-stony non-stony	prismatic or blocky	slow slow	plastic plastic	dark yellowish brown
Yellow Soil	A	colluvium	clayey silt	variable	-----	moderate	slightly plastic	yellowish brown
Sandstone Soil	---	----	sandy loam	non-stony	granular	rapid	non-plastic	----

Table 3.2

TABLE 3.2 Soil chemical properties (after Atkinson 1967)

Soil Development	Horizon	pH	base saturation %	electrical conductivity mmhos	organic matter %	CaCO₃ %	Available me / 100g P₂O₅	Available me / 100g K₂O	Exchangeable me/l Ca, Mg, Na, K
Red Mediterranean Soil	A	7.5	51	0.3	0.7	24	28	112	5.3
	B	7.6	55	0.2	0.6	10	46	31	3.0
	C	7.6	60	0.1	0.1	7	52	24	3.1
Yellow Mediterranean Soil	A	7.1	42	0.4	2.2	18	14	58	4.3
	B	7.9	43	0.3	0.7	28	50	16	2.6
Yellow Soil	A	7.8	43	0.8	3.0	18	51	21	6.7
Sandstone Soil	---	7.9	low	0.2	1.0	0	low	low	10.4

Plate 3.23 Cultivated Red Mediterranean Soil

Influences of Soils on Man

The Red Mediterranean Soils of the Transjordanian Plateau and the highlands at the edge of the Jordan River Valley have many characteristics highly suited for agriculture (pl. 3.23). These include: (1) ability to retain water for plant use; (2) a relatively high nutrient status; (3) favorable topography; and (4) low coarse fragment contents. The Yellow Mediterranean Soils, being in a drier climate and often in areas of rougher topography, are not as suited for agriculture as are the Red Mediterranean Soils. Despite their often being shallow and consequently droughty, plus having higher coarse fragment contents, the Yellow Mediterranean Soils are extensively used for field crops and grazing in the Tell Hesban project area. The calcareousness, salinity, arid climate, rough topography, and generally lower water retention capacity of the Yellow Soils generally render them unsuitable for any use except extensive grazing. The Regosolic Soils have similar limitations. The Alluvial Soils, although not having the best physical and chemical properties for agriculture, are intensively utilized in locations where perennially flowing water is available.

Influences of Man on Soils

Man has had a marked influence on the soils of the project area in that he has converted soils of a steppe-type of environment to cultivated soils and soils typical of heavily utilized rangeland. Thus, the soil's physical and chemical properties have been altered, generally for the worse. As discussed previously, it is also very probable that man's settlement, cultivation, and grazing activities have accelerated the rates of soil erosion in the area. However, on the positive side, introduction of soil conservation practices, at present, and probably also during some of the more progressive periods of history, is rehabilitating and reclaiming some areas that were previously barren. Thus, with the current increase in farming in the area, there may be a net accretion to the already considerable soil resources of the area.

References

Amiran, D. H. K., and Gilead, M.
1954   Early Excessive Rainfall and Soil Erosion in Israel. *Israel Exploration Journal* 4: 286-295.

Atkinson, K. et al.
1967   *Soil Conservation Survey of Wadi Shuieb and Wadi Kufrein, Jordan.* University of Durnam.

Bender, F.
1975   *Geology of Jordan. Contributions to the Regional Geology of the Earth.* Dr. H. J. Martini ed. Supplementary Edition of Volume 7, Berlin: Gebuder Bortraeger.

Bullard, R. G.
1972   Geological Study of the Hesban Area. *Andrews University Seminary Studies* 10: 129-141.

Butzer, K. W.
1961   Paleoclimatic Implications of Pleistocene Stratigraphy in the Mediterranean Area. *Annals of the New York Academy of Science* 95: 440-456.

Dan, J.
1977   The Distribution and Origin of Nari and Other Lime Crusts in Israel. *Israel Journal of Earth Sciences* 26: 68-83.

Fleming, R. W., and Johnson, A. M.
1975   Rates of Seasonal Creep of Silty Clay Soil. *Quarterly Journal of Engineering Geology* 8: 1-29.

Food and Agriculture Organization of the United Nations
1974   *Soil Map of the World.*

Horowitz, A.
1979   *The Quaternary in Israel.* New York: Institute of Archaeology Academic Press.

James, H. E., Jr.
- 1976 Geological Study at Tell Hesban. *Andrews University Seminary Studies* 14: 165-169.

Moorman, E.
- 1959 *The Soils of East Jordan: Report to the Government of Jordan.* Expanded Technical Assistance Program No. 1132. Rome: Food and Agricultural Organization of the United Nation.

Soil Survey Staff
- 1975 *Soil Taxonomy, A Basic System of Soil Classification for Making and Interpreting Soil Surveys.* Soil Conservation Service, United States Department of Agriculture, Agricultural Handbook No. 436.

Stockton, E. D.
- 1969 A Bibliography of the Flint Industries of Jordan. Pp. 100-103 in *Levant*, the British School of Archaeology in Jerusalem, London.

Thompson, W. M.
- 1880 *The Land and the Book*, Vol 3. New York: Harper and Brothers.

United States Department of Agriculture
- 1938 *Soils and Men.* Yearbook of Agriculture 1938, United States Government Printing Office pp. 979-1001.

West, B. G.
- 1970 *Soil Survey in the Baq'a Valley.* Consultant Report 4, Food and Agriculture Organization of the United Nations.

Yaalon, D.
- 1975 *Some Data on the Nature and Origin of Nari (Calcrete) on Chalk in Israel.* Proc. Colloque Int. sur Types de Croutes Calcaires et Leur Repartition Regionale, Strasbourg, pp. 12-13.

Chapter Four
SURFACE AND GROUNDWATER RESOURCES OF TELL HESBAN AND AREA, JORDAN

Larry Lacelle

Chapter Four
Surface and Groundwater Resources of Tell Hesban and Area, Jordan

Introduction

This paper discusses the surface water resources of the Tell Hesban project area in terms of the springs in the project area, their rates of flow, and their water quality. The groundwater of the area is discussed in terms of the nature of the aquifers, the quantities of groundwater, its accessibility, and its quality. Only those aspects of surface water and groundwater studies that affect or have affected man's use of the land are discussed. More complete technical evaluations of the surface and groundwater resources of Jordan are given in: Agrar und Hydrotechnik (1977), Humphries (1978), Hydrology Division (1966), Ionides and Blake (1939), MacDonald et al. (1965), and Manners (1969). As the Agrar and Hydrotechnik study "National Water Master Plan for Jordan" consists of a massive re-evaluation of all previous data, plus presentation of much new and pertinent data, it is the primary reference used.

Surface Water

The Tell Hesban project area, encompassing portions of the Transjordanian Plateau, the highlands (hills) at its western edge, and the wadis leading down to the Jordan River and Dead Sea, is neither one complete drainage basin, nor one complete system of watersheds (fig. 4.1). The project area's northern and western portions are part of the lower Jordan River Basin, while its southern and eastern portions are part of the Dead Sea Basin. The western portion of the project area consists primarily of three main watercourses that drain the steep escarpment of the Jordan River Valley. These are the headwaters and midstream portions of the watersheds wadis Hesban, Muhtariqa, and Kanisa (fig. 4.1). Wadis Hesban and Muhtariqua flow into the Jordan River to the west of the project area, whereas Wadi Kanisa drains into the Dead Sea. The northernmost portion of the project area, the hills of Naur, is part of the headwaters of the Wadi Nusariyat watershed, nearly all of which is outside of the project area. All of the above mentioned westward draining wadis are deeply incised into the edge of the plateau and all have spring fed perennial flow.

In contrast to the wadis draining west, the eastern portion of the project area on the plateau is drained to the south by a number of small wadis (Habis, Manshiya, and Amad) that are tributaries to Wadi Zerqa Main. These wadis consist of irregularly sinuous, intermittently flowing streams in gently sloping valleys on the plateau surface. They are not deeply incised within the project area.

Springs, and the streamflow resulting from them, are the only perennial surface water resources available for man's use in the Tell Hesban project area. Neither natural surface water bodies, nor rain fed perennially flowing streams exist on the surface of the Transjordanian Plateau or in the highlands along its western edge. This lack of surface waters is due to factors such as: semiarid climate, soils with a high ability to infiltrate and store precipitation, and relatively permeable carbonate bedrocks. However, in order to accommodate his water needs, man has, in the past, constructed numerous small, and several quite large, cisterns and reservoirs at Tell Hesban and Madaba (Boraas and Geraty 1976: 107-108, 1978: 12-13; Thompson 1880: 637; Tristram 1865: 540, Tristram 1873: 326) (pl. 4.1). These manmade surface water bodies were constructed with systems of channels and conduits designed to channel surface waterflow into them during the rainy season (pl. 4.2).

Wadis have incised themselves deeply into the western edge of the Transjordanian Plateau, and in doing so have exposed extensive areas of the primarily limestone (upper) aquifers and the

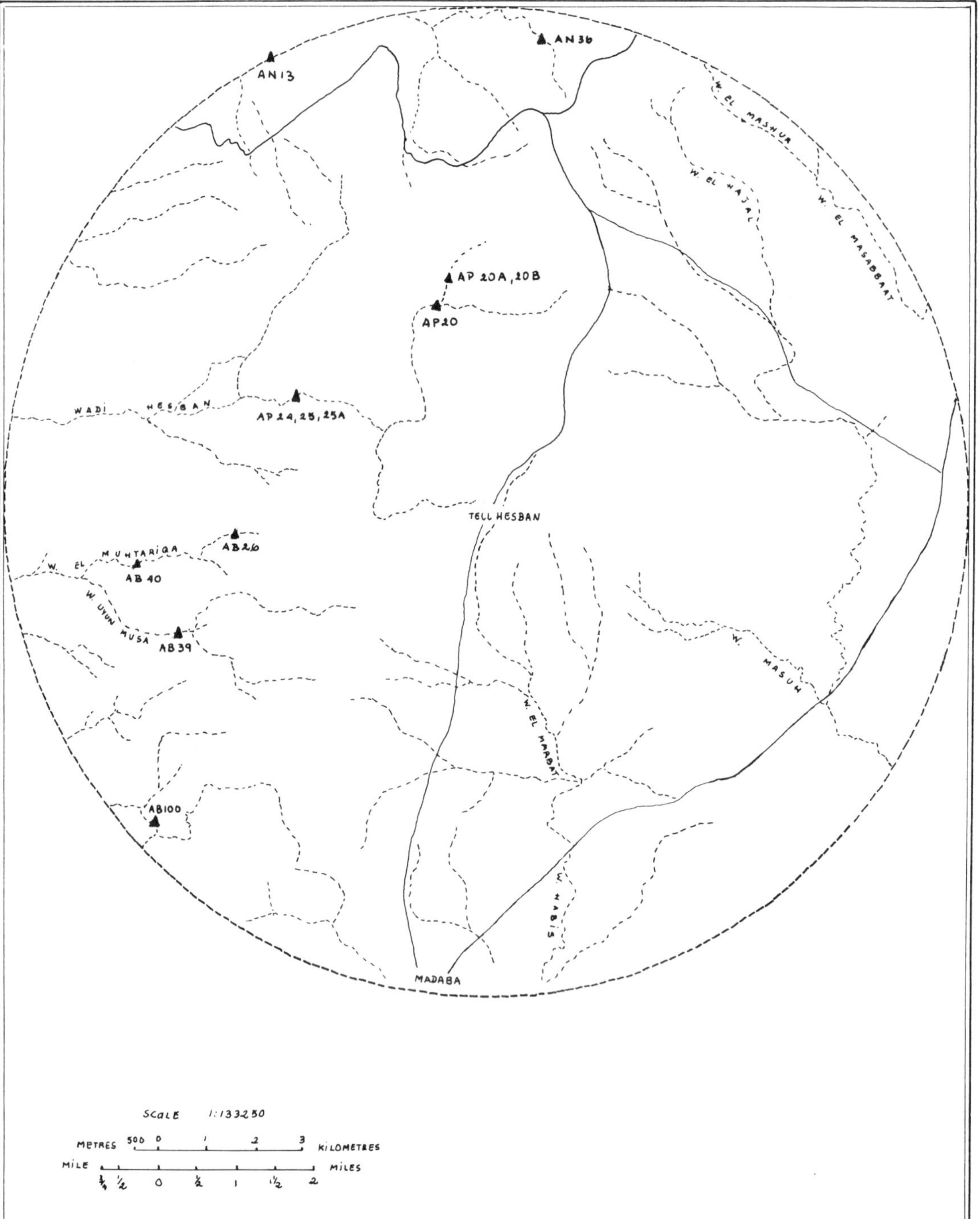

FIGURE 4.1 Major drainages and locations of springs in the Tell Hesban project area (see Tables 4.1 and 4.2 for spring names)

TID

SURFACE AND GROUNDWATER RESOURCES 63

Plate 4.1 Large ancient reservoir at Tell Hesban

Plate 4.2 Cistern in depression with channels to collect overland flow during rainstorms

primarily sandstone (lower) aquifers that underlie the plateau. At some 18 locations in the project area, there are perennially flowing springs emitting from these aquifers (fig. 4.1). The outflow of the aquifers into the watercourses is described as the baseflow (groundwater runoff) for the area. Baseflow is greatest in the winter in response to rainfall replenishment of the aquifers. It generally decreases, or, in some smaller springs, stops in the dry summer season.

Another major, although temporary, surface water resource in the project area is floodflow (surface runoff). In this area, floodflow only occurs in the streambeds during, or immediately following storms that are intensive enough to exceed the ability of the soil and vegetation to absorb the rainfall. Examination of stereoscopic aerial photography suggests that in the small catchments on the plateau surface and in the hills, some dams and waterworks may have been constructed in the past in order to confine floodflow for irrigation or domestic uses. The remains of one dam were observed near Ain Hesban, but, being in an irrigation area, this dam may have been a divisionary structure for irrigating, rather than a reservoir for floodflow (pl. 4.3). Whether other dams or waterworks were constructed in the past in the wadis of the project area, is certainly a question worth investigating, as the study of such structures would result in valuable insights as to the degree of sophistication of ancient water management.

As the springs in the project area are all in the wadis at lower elevations, and often difficult to reach from above, the past residents of the plateau and hills appear to have been largely independent of these water sources. Tell Hesban is likely an exception though, as the springs at Ain Hesban are sufficiently near to make water hauling a viable enterprise, even with the primitive means of transport available in the past (pl. 4.4). One of the early European travellers in this area, Thompson (1880: 664), described travelling along an ancient, well-constructed road from Tell Hesban down to Ain Hesban. The Tell Hesban and area citizens are most certainly dependent on the springs today, as tanker water deliveries are regularly used to supplement the natural precipitation inputs into the cisterns.

Irrigation systems based on use of the perennial flow from the numerous springs in the wadis are well-developed today (pl. 4.5). Ruins and ancient appearing terracing at irrigated areas near major springs, such as Ain Hesban and Ain Musa, certainly suggest that irrigation systems were well-developed in the past also. Major springs were also utilized in the past as sources of hydropower to turn waterwheels for grinding mills (Thompson 1880: 666; Tristram 1873: 356). Further research as to the extent of past irrigation development would give us valuable insights as to the degree of sophistication employed in ancient cultivation practices.

For the present, Agrar und Hydrotechnik (1977: Vol. 5, Annex 1, p. 3) report that some 932 acres (377 hectares) of fields are presently irrigated in the wadis below springs in the project area (table 4.1).

Manners (1967: 162) suggests that as little as 1/6 of the above acreages may be the actual amount irrigated in any one year, due to water shortages. Minimum and maximum volumes of baseflow from the major perennial springs in the project area are given in table 4.2. As baseflow is lowest during the dry season when irrigation is required, the minimum volumes are likely the most relevant values to use in evaluating the irrigation potential of this area. As the reported minimum values are rather low, it is probable that acreages irrigated in the past were not much different, or were less than those irrigated today. A number of smaller springs that are not mentioned in table 4.2 also are found in the project area. They often dry up during the summer and are used for local domestic and stock watering purposes only (pl. 4.6).

The Wadi Kafrein immediately north of the Tell Hesban project area has a salinity rating (total dissolved solids) of approximately 750 ppm at its outlet (Agrar und Hydrotechnik 1977: map HG 2.1). According to their rating scheme, water of this salinity level is fair for water supply, good to medium for stock watering, but has a high salinity hazard. Four major springs in Wadi Hesban have total dissolved solids of 320 to 360 ppm, a value that rates them as good for irrigation use (Humphries 1978: 230).

The volume of sediment transported in the wadis on the Jordan River Valley escarpment does not appear to be excessively large in terms of sediment or bedload. Based on earlier studies, Agrar und Hydrotechmik (1977: Vol. 3, p. 45) suggest that suspended sediment loads of approximately 3% of total floodflow from runoff, and bedloads of approximately 30% of annual

SURFACE AND GROUNDWATER RESOURCES 65

Plate 4.3 Remains of an old dam near Ain Hesban

Table 4.1

TABLE 4.1 Irrigated acreage in the Tell Hesban project area

Watershed Reference	Irrigation Area #	Irrigation Area Name	Irrigated Area ac.	ha.
AN	13	Kufrien	228.5	92.5
AN	36	Naur	49.4	20.0
AP	37	Er Routah	7.4	3.0
AP	38	Wadi Hisban	481.6	195.0
AB28	40	Ain el Jammala	17.3	7.0
AB28	39	Ain Musa (N&S)	148.2	60.0
			932.4	377.5

66 ENVIRONMENTAL FOUNDATIONS

Plate 4.4   Illustration of water hauling by donkey, representative of ancient methods used at Tell Hesban

Table 4.2

TABLE 4.2 Maximum and minimum flows for major springs in the Tell Hesban project area (after Hydrology Division, 1966)

Watershed Reference	Spring #	Name	Flows (liters/sec.) max.	min.
AN	36?	El Kubia	1.7	0.2
AN	36?	El Khashabeh	0.8	0.1
AN	36?	El Jami	0.5	0.0
AN	36?	Naur (El Balad)	4.7	0.1
AN	36?	Nasara	15.8	0.5
AN		Um Zweiteeneh	6.8	1.6
AN		Adasyia Gharbiya	9.9	0.2
AP	20	Ain Hisban	200.0	52.0
AP	20A	Faria	108.0	31.0
AP	20B	Kdeeshah	25.6	1.7
AP	24	Sumiya	26.3	15.3
AP	25	Mdeirdeh	70.6	19.3
AP	25A	Mazzazat	20.0	14.6
AB26	39	Musa north	11.0	3.8
AB26	39	Musa south	15.0	0.3
AB26	100?	Ain Judeid		
AB26		Ain el Beida		
AB	40	Jammala	1.2	trace

Plate 4.5 General view of an area irrigated by water systems from a perennial spring

suspended load, are appropriate rough estimates. As most sediment transport occurs during winter storms, suspended sediment/bedload does not have a major impact on the quality of irrigation/domestic water, which is mostly used in the summer.

Total annual average streamflow out of wadis Hesban, Muhtariqa, and Kanisa (the major drainage basins in the project area having perennial flow) (fig. 4.1), is approximately 7.9 million cubic meters (MCM), a figure which does not include some 2.9 MCM of water that is used upstream for irrigation (Agrar und Hydrotechnik 1977: Vol. 3, Annex 3A-3.6, p. 5). Dry year outflow values are given as approximately 6.7 MCM outflow, and wet years 9.7 MCM. For Wadi Hesban, approximately 90% of this outflow is baseflow from springs and 10% floodflow. The percentage of floodflow is reported to vary between 0 to 20% for abnormally dry, or wet years. Outflow data for the wadis Muhtariqa and Kanisa suggest that most floodflow off uplands is reabsorbed by the soils or bedrock aquifers further downstream. For average or dry years these wadis have virtually no floodflow measured at their outlets and only 7% in wet years. The quantities of outflow from the major watersheds in the project area indicate that there is still some opportunity for further water apportionment for the development of agriculture or domestic use.

Influences of Surface Water Resources on Man

The lack of natural surface water resources on the plateau has undoubtably restricted man's use of the land there. However, by careful conservation of rainfall and water hauling, coupled with soils capable of retaining sufficient water to nurture field crops and range plants into the dry season, man has been able to live on and farm this area quite successfully. In the wadis, the springs have been, and are, a focus for irrigation farming. This resource has not been further expanded, probably because the rough topography, extensive exposed bedrock, and shallow, often coarse textured soils (see the Geology, Surficial Geology, and Soils chapter in this volume), are major constraints on further development.

Influences of Man on the Surface Water Resources

Man's primary influence on the surface water resources has been to try to confine and divert it to try to meet his domestic and irrigation needs. The inconvenient location of most of the major springs in the steep walled wadis has challenged man's ability to harness them and to modify the rugged slopes to make them sufficiently arable to be worth irrigating. Man does not appear to have deteriorated the surface water resource, as large quantities of good quality water still flow westward out of the project area. In fact, his construction of storage facilities, from small cisterns to wadi dams, has enhanced the resource by vastly increasing the utilization of the winter floodflow. The influences of man on surface water and surface water on man are further discussed in Chapter Eight of this volume.

Groundwater

Groundwater resources, as evaluated in this section, include a brief description of water transport and storage in the soil, as well as the evaluation of quantities and quality of the groundwater in the aquifers beneath the Tell Hesban project area.

The soils are evaluated as part of the groundwater resource in this section as they have a very significant role in infiltrating and retaining precipitation in a form available for later plant use, as well as acting as a conduit for recharge of the groundwater aquifer. Atkinson *et al.* (1967) estimate a soil infiltration rate of 14 mm per hour for the wadis Shueib and Kafein immediately north of the project area, this rate likely being approximate for the Tell Hesban project area also. They observe that with maximum precipitation intensities on the order of 17 mm/hr, there should be little runoff on level land. MacDonald *et al.* (1965) estimate that 4 to 8% of a heavy precipitation will flow over the soil surface and be discharged from the immediate area. Labadi (1959) estimates the runoff to be 5 to 15%. Thus, up to 15% of a heavy rainfall has no opportunity to recharge the soil or groundwater aquifer at that location. However, the runoff may be reabsorbed downstream, as previously discussed.

When the entire soil depth has been recharged with water, it can begin to drain downward into the bedrock aquifer. However, if the rainfall is insufficient to fully recharge the soil, all the infiltrated water will remain in the soil to be used by plants, or lost to evaporation. Agrar und Hydrotechnik (1977: Vol. 3, p. 39) suggest that an average amount of soil moisture available for evapotranspiration (evaporation plus plant transpiration) in the Wadi Hesban area is approximately 203 mm per year, whereas the potential evapotranspiration (the amount that could be used if an abundance of water were available) is 1639 mm per year. This large soil water deficit (-1436 mm) emphatically illustrates that much more soil water could be used than is available. Thus, it becomes apparent that large recharges through the soil to the bedrock aquifers are not likely.

Despite the above observations, some rainfall does pass through the soil to the bedrock, and some is directly absorbed through porous bedrock surfaces, fractures, and solution pockets on bedrock exposures (of which there are many in this area). Thus, Ionides and Blake (1939) estimate that for the watersheds in this area that approximately 5.3% of total precipitation inputted to a watershed is returned as springflow in the wadis. Agrar und Hydrotechnik (1978: Vol. 1, Annex 3, p. 2) also report annual natural recharges to the bedrock aquifers of the wadis Ḥesbân, Muhtariqa, and Kanisa watersheds (fig. 4.1) of some 12.3 million cubic meters (MCM) per annum. In evaluating the relative quantities of recharge and outputs, these authors (Vol. 1, Annex 7, p. 4) suggest that there is some 3.0 MCM per annum of groundwater available in the aquifers supplying the above mentioned wadis.

Unfortunately, the 3.0 MCM of groundwater surplus is not easily accessible from the plateau surface. Agrar und Hydrotechnik's map (HG 5.4) depicts a depth of greater than 150 m to the saturated aquifer, a depth usually considered to be too deep to practically/economically pump from. Three wells (PP 83, PP 313, and W 73) on the eastern fringes of the project area have depths to water of 166, 150, and 137 m, respectively (Agrar und Hydrotechnik 1977: Vol. 4, List 1.1, pp. 8, 11, 18) (fig. 4.2). Even at these depths, flow volumes are weak, approximately 1 1/2 cubic meters per hour for PP 83 and PP 313 (Ionides and Blake 1939), and 20 M^3/hr for W73 (Agrar und Hydrotechnik 1977: Vol. 4, List 1.1, p. 18). Thus, for modern man, and especially for ancient man, obtaining water from wells on the surface of the plateau is not practical. However, where the wadis of the Jordan River Valley have incised into the plateau, well depths to saturated aquifers are much less. Agrar und Hydrotechnik (map L1-4N) depict extraction of groundwater from wells for irrigation in both wadis Hesban and Naur in the project area, the wells likely being near the springs.

Agrar und Hydrotechnik (1977: Vol. 4, pp. 2-15) and Humphries (1978: 224-227) describe the bedrock types, and the physical and chemical characteristics of the groundwater aquifers underlying the project area. The upper aquifer consists primarily of carbonate rocks (upper Belqa group and lower Ajlun group of Upper Cretaceous age) (fig. 4.2). These rocks are primarily limestones, marls, and chalks, with some chert and shale strata. The Belqa and Ajlun groups vary from some 500 to 1000 m thick (Humphries 1978: 226-227) under the plateau. Bedrock permeabilities and rates of water transmisivity are reported to be quite variable depending upon rock type (chalks and marls are relatively permeable), with the degree of faulting/fracturing, and with the dip of the strata. Depth to the saturated aquifer also varies with the above. Wells hitting a fractured zone in otherwise impermeable bedrock might be good producers, whereas nearby wells in solid rock might be dry. Springs such as Ain Hesban occur where the wadis have cut into zones of high transmisivity in the aquifer.

As the bedrock of this aquifer is calcareous, the primary contaminant in the groundwater is calcium. When a greater proportion of sodium and magnesium is present, the water becomes more saline and much less suitable for use. Agrar und Hydrotechnik (1977: map HG 2.1) suggest that the upper reservoir groundwater has a salinity of from 500 to 1000 parts per million (PPM) total dissolved solids (TDS), and an electrical conductivity of 750 to 1500 micromhos (mmhos). According to their rating system, the water of this aquifer is, on the average, fair for water supply, high for salinity hazard and good to medium for stock watering. Agrar und Hydrotechnik (1977: map HG 2.1) suggest that in the upper northeast corner of

SURFACE AND GROUNDWATER RESOURCES 71

Fig. 4.2 Geological cross-section near Tell Hesban (after Agrar und Hydrotechnik, 1977, Map HG 1.3)

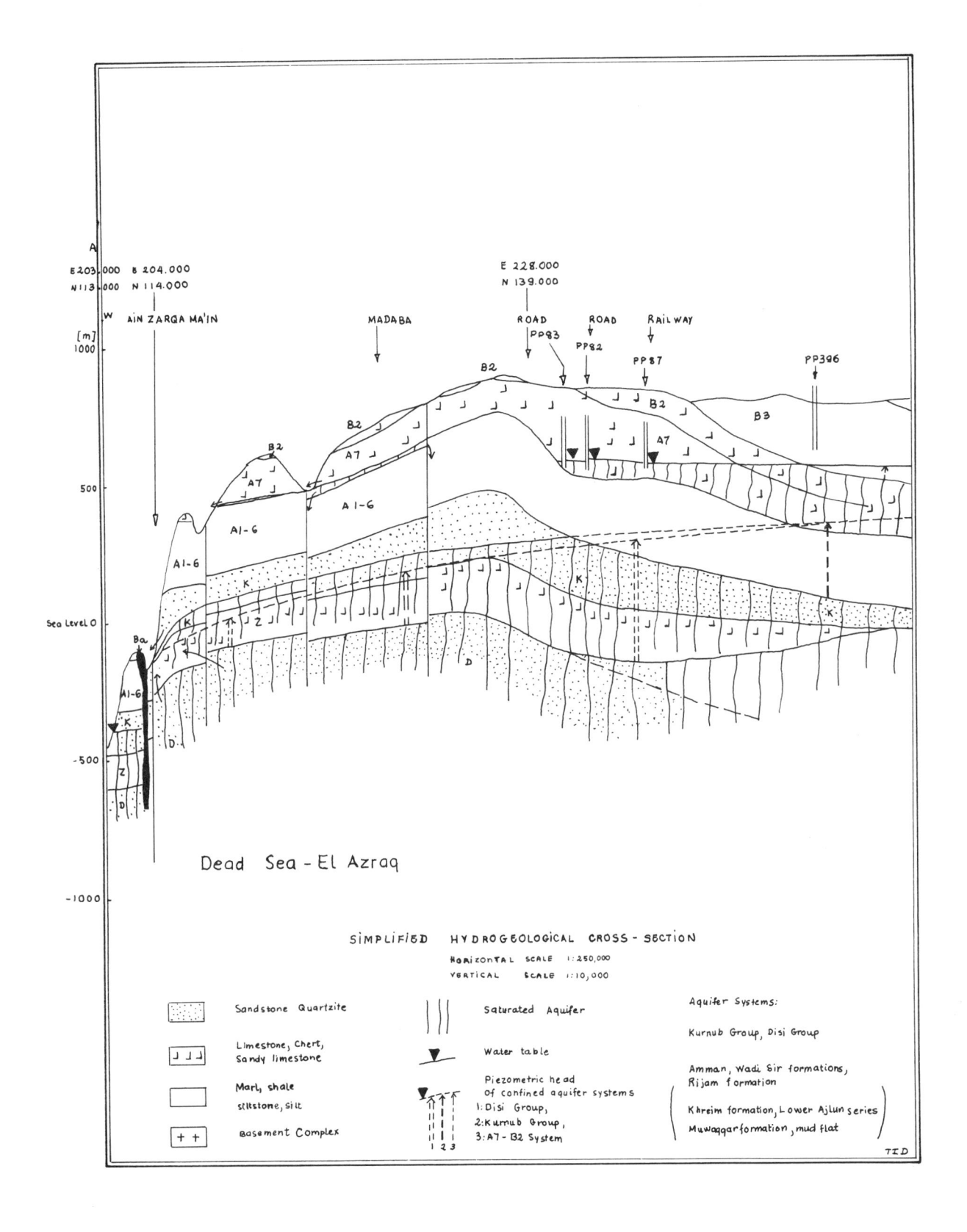

the project area, the upper aquifer is contaminated by pollutants from the Amman area.

The lower aquifer (the Kurnub group of Lower Cretaceous age) (fig. 4.2) consists primarily of sandstones and marls, and is some 350 m thick (Humphries 1978: 225). These rocks are a good aquifer in that they are generally permeable and able to transmit groundwater. As this aquifer lies beneath the great thicknesses of the Ajlun and Ajlun groups, it is effectively inaccessible from the surface of the plateau. However, as it does outcrop in the westernmost and lowest part of the project area, it is locally suitable for wells. Several major springs in the wadis of the project area occur due to waterflow along the zone where the Ajlun and Kurnub bedrocks meet.

Agrar und Hydrotechnik (1977: Vol. 4, p. 14) suggest that, like the upper aquifers, the salinity hazard in Kurnub group aquifer is medium to high. It is rated fair for drinking water and good to medium for stock watering.

Influences of the Groundwater
Resources on Man

The above data clearly show that there is a significant amount of groundwater available under the plateau for man's use. Unfortunately for those on the plateau, it is too deeply buried to be extracted by wells. However, in the wadis, the potential for wells of a reasonable depth is much better, especially in the vicinity of springs. On the plateau, it is really only the large capacity that the local soils have for absorbing water that permits adequate crop and range plant growth for man to successfully farm in such a water deficient environment.

Influences of Man on the Groundwater
Resources

Except for a limited area of subsurface aquifer pollution from the Amman metropolitan area, man has had virtually no effect on the groundwater reserves of the project area. Even the wells in the irrigated areas of the wadis do not appear to have significantly drawn down the aquifer and thus curtailed flow from the springs. However, extensive future well-development in the vicinity of the springs could certainly damage the very important springwater resources.

The influences of groundwater on man and man on groundwater are further discussed in Chapter Eight of this volume.

References

Agrar und Hydrotechnik GMBH and Bundesanstalt Für Geowissenschaften und Rohstoffe.
  1977 *National Water Master Plan for Jordan*. The Hashemite Kingdom of Jordan, Natural Resources Authority, Amman and the Federal Republic of Germany, German Agency for Technical Cooperation Ltd., Frankfurt, 6 volumes plus maps.

Atkinson, K. *et al.*
  1967 *Soil conservation Survey of Wadi Shueib and Wadi Kafrein, Jordan*. University of Durnam.

Boraas, R. S., and Geraty, L. T.
  1976 *Heshbon 1974, The Fourth Campaign at Tell Hesban, A Preliminary Report*. Berrien Springs, MI: Andrews University Press.

  1978 *Heshbon 1976, The Fifth Campaign at Tell Hesban, A Preliminary Report*. Berrien Springs, MI: Andrews University Press.

Humphries, Howard & Sons
  1978 *Water Use Strategy: North Jordan Vol. 2, Water Resource*. Reading, England: Howard Humphries & Sons Consulting Engineers, pp. 5-50, 223-238.

Hydrology Division
  1966 Review of Spring Flow Data Prior to Oct. 1975. Natural Resources Authority, Department of Research and Investigation, Hydrology Division, Technical Paper No. 40, pp. 57-59.

Ionides, M. G., and Blake, G. S.
  1939 *Report on the Water Resources of Transjordan and Their Development*. Published on behalf of the Government of Jordan by the Crown Agents for the Colonies, 4 Millbank, London, SW 1.

Labadi, A. M.
1959 *Surface Water Hydrology of Jordan.*

MacDonald, Sir M. and Partners
1965 *Hydrological Survey of the Madeb-a-Man Area.* The Hashemite Kingdom of Jordan, Central Water Authority, Vol. 3, Groundwater Recharge.

Manners, I. R.
1969 *The Development of Irrigation Agriculture in the Hashemite Kingdom of Jordan, With Particular Reference to the Jordan Valley.* Ph.D. Thesis, Oxford: Bodelian Library, pp. 100-119, 168-174, 196-199.

Thompson, W. M.
1880 *The Land and the Book*, Vol. 3. New York: Harper and Bros.

Tristram, H. B.
1865 *The Land of Israel, A Journal of Travel in Palestine.* London: Society for Promoting Christian Knowledge.

1873 *The Land of Moab.* New York: Harper and Bros., Publishers.

Chapter Five
FLORA OF TELL HESBAN AND AREA, JORDAN

Patricia Crawford

Chapter Five
Flora of Tell Hesban and Area, Jordan

Introduction

This paper describes the flora of the Tell Hesban project area (fig. 1.1) in terms of the observed species; their descriptions, habitats, and economic uses. Related ecosystem data on the flora of the project area are contained in Chapter Six, "Ecology of the Flora of Tell Hesban and Area, Jordan." Relevant data on past flora are contained in Chapter Seven, "Paleoethnobotany and Paleoenvironment of Tell Hesban and Area, Jordan."

The study of the flora of the Tell Hesban project area has been undertaken in three stages. The first stage consisted of gathering plant specimens from unspecified localities during the 1974 excavation season. These specimens were gathered without supporting data and therefore were not useful for purposes of discussion of ecological relationships. Nevertheless, this material provided a preliminary view of the diversity of flora, both natural and cultivated, which exist in the area (Crawford and LaBianca 1976).

The second period of data gathering occurred during the excavation season in 1976 and was undertaken by myself in a systematic way to provide a more controlled picture of the various plant-soil-habitat relationships. This collection of data was limited to the immediate vicinity of the tell and thus did not provide a picture of the overall patterns of the whole project area. It did provide a controlled sampling of the type of flora indicative of disturbed areas, as well as providing a baseline for patterns of vegetation to be observed elsewhere.

The third stage of data collection occurred over a two-week period in July 1979 and involved surveying the entire Tell Hesban project area, a 10-km radius centered at Tell Hesban (fig. 1.1). The survey work during these two weeks was also integral in preparation of Chapter Six of this volume, "Ecology of the Flora of Tell Hesban and Area, Jordan." During this brief field expedition, numerous walking transects were carried out throughout the extent of the project area. Unknown species were collected for later identification, and known species were documented at each site investigated. Notes were also taken on soil, drainage aspect, and landuse practices at each site. A soil scientist gathering data for the "Ecology of the Flora . . ." chapter was present during the survey period. By combining our respective fields of expertise, it was possible to gain a good understanding of the environmental factors contributing to, and accounting for, the various patterns of vegetation observed. Specimens collected were identified with the assistance of Dr. Peter Stevens and Dr. Uzi Plitmann of Harvard University, and Dr. Loutfy Boulos, at that time at the University of Jordan, presently at Cairo University.

The completeness of the data reported in this paper is limited by the fact that all three collection periods were during the summer drought, the time of year when many of the characteristic flora are wilted and decayed. It is also a time of year when the entire project area is being heavily grazed, thus tending to limit the number of succulent species available for collection and identification.

The various types of ecological niches surveyed within the Tell Hesban project area include: (1) the tell and disturbed areas immediately adjacent to it; (2) domestic gardens and yards; (3) road and path sides; (4) harvested wheat and barley fields, and cultivated vegetable fields; (5) olive groves; (6) terraced, cultivated areas—usually grapes; (7) dried-up wadi beds; (8) banks of perennially flowing streams; (9) springs; (10) ancient cultivation terraces; (11) the course of an ancient Roman road; (12) rangeland areas on steep, rough topography; and (13) reforested coniferous groves. It is assumed, however, that every one of the above mentioned areas has been exposed in the past, and is currently being exposed to constant grazing by sheep and goats. Goats appear to be capable of grazing even the roughest terrain in the project area and are the least fussy about what they eat.

Fig. 5.1 The Tell Hesban project area

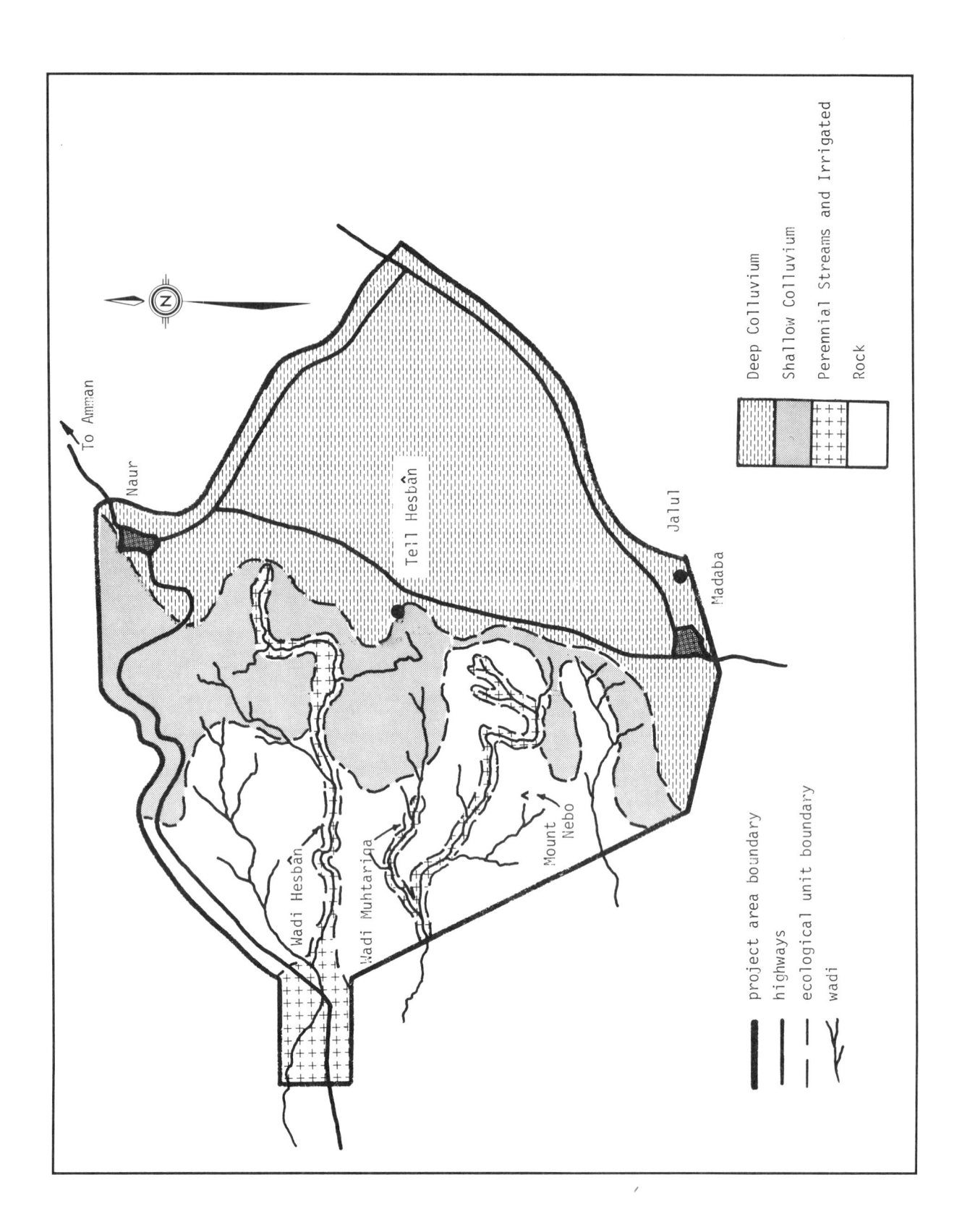

Of the areas traversed, those supplied with a continuous source of water afforded the greatest variety of flora. Recently harvested cereal crop fields had the least number of species. The kitchen gardens afforded the greatest number and concentration of domesticated species.

The Species Descriptions section of this chapter lists all the observed flora in the project area. It also describes the physical characteristics of the flora, their habitats, and their economic uses. A list of cultivars observed growing in the project area is given in the Cultivated Species section, so that a more complete picture of the landuse and potential can be drawn. The uses and the characteristics of these plants are not discussed, however, as cultivars represent species, both native and introduced, which inhabit and have been adapted to the local environment through the conscious manipulations of man. The Species List section of this chapter is an alphabetical list by Latin name of all species observed in the project area. The page number relating to the description of each species in the Species Description Section is also given for convenience.

Economic Uses of Plants

The description of flora in the Species Description section lists documented, or observed, economic, and grazing uses of various plant species. This data, plus lists of species and their uses in Zohary (1973: 612-617), have been utilized in this section to generally illustrate the ranges of uses of the local flora. It should be emphasized, however, that documentation of a use does not necessarily mean that that particular species has been so used in the Tell Hesban project area. Only the potential for use is indicated, unless the use was actually observed. Observed uses are documented in the text of the Species Description section. Further data on past species found at Tell Hesban and their possible uses is contained in Chapter Seven of this volume. The final part of the Economic Uses of Plants section briefly lists many of the most commonly observed species of the area that are known to be suitable for grazing and/or for fodder crops.

Plants Utilized as Food

Vegetables

A number of local plant species, many appearing to be rather unpalatable for humans, may, in fact, be utilized as cooked vegetables. Although the mature plant may be spiny and tough, the young shoots and flower heads of many species are succulent, nutritious, and tasty. Species described in the Species Descriptions section as potential sources of vegetables include: pigweed (*Amaranthus*), saltwort (*Atriplex*), common goosefoot (*Chenopodium*), centaury (*Centaurea*), chicory (*Chichorum*), Holy thistle (*Silybum*), sharp leaved asparagus (*Asparagus*), mallow (including okra) (*Malva*), black nightshade (*Solanum*), and blue eryngo (*Eryngium*).

Salads

As with species utilized for vegetables, it is generally the young shoots and leaves that are used for salad greens. Species described in the Species Descriptions section that may be used in salads include: chicory (*Chichorum*), Holy thistle (*Silybum*), sow thistle (*Sonchus*), and dock (*Rumex*).

Bulbs and Roots Species

Local plants having roots that may be utilized as food include: cattail (*Typha*), camel thorn (*Alhagi*), and storksbill (*Erodium*).

Wild Fruits and Nuts

Quite a number of the local shrub and tree species bear fruits that are, or could be, part of the local diet. Plants producing edible fruits include: prickly pear cactus (*Opuntia*), caper (*Capparis*), fig (*Ficus*), mulberry fig (*Ficus*), olive (*Olea*), pistachio (*Pistacia*), jujube (*Zizyphus*), hawthorne (*Crataegus*), and *Lycium*. Almond (*Prunus*) produces a nutritious nut. Olive and almond also produce valuable oils.

Spices

Local species which may be utilized as spices include: sage (*Salvia*), sagebrush (*Artemisia*), mustard (*Brassica*), mint (*Mentha*), and headed savory (*Thymus*).

Medicinal Uses

A large number of the local species have potential for use as herbal medicines. Those that are known to be cathartic or purgative include pepper tree (*Schinus*), field bindweed (*Convolvulus*), bitter gourd (*Citrullus*), spurge (*Euphorbia*), and fig (*Ficus*). Species that may be used to induce urination (diuretics) include mulberry fig (*Ficus*) and common peganum (*Peganum*). Species that may be used to induce vomiting include turnsole (*Chrozophora*) and common peganum (*Peganum*). Species utilized as stimulants or tonics include common chicory (*Marrubium*). For respiratory ailments, Saint John's wort (*Hypericum*) (inhale steam) and white horehound (*Marrubium*) (for coughs), may be utilized. Species potentially useful as blood thinners include milfoil (*Achillea*), yarrow (*Achillea*), and sweet clover (*Melilotus*). Plant species reportedly suitable for poultices include fig (*Ficus*) and knotweed (*Polygonum*). Other potential medicinal species from the Species Description section include: Oleander (*Nerium*) (an active cardiac glucoside), sagebrush (*Artemisia*), thistle (*Centaurea*) (for ague and jaundice), sow thistle (*Sonchus*) (an esculent), headed savory (*Thymus*), and larkspur (*Delphinium*) (a pain reliever). In addition, Zohary (1967: 614) suggests that a number of additional species found in the Tell Hesban area also have medicinal uses. They are camel thorn (*Alhagi*), caper (*Capparis*), Bermuda grass (*Cynodon*), mint (*Mentha*), sage (*Salvia*), and small caltrops (*Tribulus*).

Plants Utilized for Industrial Purposes

Resin, pitch and tannin may be extracted from Aleppo pine (*Pinus*) and gum from pepper trees (*Schinus*). Oils for perfume may be extracted from headed savory (*Thymus*). Zohary (1973: 615) suggests that commercial extractants also may be obtained from mustard (*Brassica*) and mint (*Mentha*).

Dyes

Two local species may be utilized as dyes. They are turnsole (*Chrozophora*) and mullein (*Verbascum*).

Chemicals

Insecticides may be extracted from two local species, Oleander (*Nerium*) and larkspur (*Delphinium*), while extractants from sage (*Salvia*) are reportedly useful for fumigation.

Plants Utilized for Domestic Purposes

Fuel and Timber

Although large trees are not common in the project area, a number of local species represent valued timber sources. They include: Australian oak (*Casuarina*), Mediterranean cypress (*Cypressus*), mulberry fig (*Ficus*), and aleppo pine (*Pinus*). Shrubs and trees commonly utilized for fuel wood include: camel thorn (*Alhagi*), aleppo pine (*Pinus*), and thorny burnet (*Poterium*).

Fences and Hedges

Thorny saltwort (*Noaea*) and thorny burnet (*Petrium*) are reportedly often used to make dry shrub fences. Prickly pear cactus (*Opuntia*), Australian oak (*Casuarina*), Mediterranean cypress (*Cupressus*), and *Lycium* are all species reportedly suitable for use as hedges or fences.

Shade, Ornamental, and Sacred Trees

Isolated trees in farmer's fields, or trees around houses are important sources of shade in the hot summers. Local species reportedly appropriate for this use include pepper tree (*Schinus*), Mediterranean cypress (*Cupressus*), jujube (*Zizyphus*), tamarisk (*Tamarix*), and pistachio (*Pistacia*). Isolated, often decorated, sacred trees of various species are locally seen in this area. One such tree, a jujube (*Zizyphus*), was seen in the project area.

Household Uses

Two local species when crushed, produce aromatic material suitable for deodorizing floors. They are headed savory (*Thymus*) and sweet clover (*Melilotus*). The common reed (*Phragmites*) that grows along the watercourses in the project area, may be effectively utilized to make mats or thatching.

Plants Utilized for Fodder and Forage

A number of the species of the local flora can also be cultivated as fodder or pasture plants. They include: wild oats (*Avena*), brome grass (*Bromus*), wild sorghum (*Sorghum*), Medick (*Medicago*), and sweet clover (*Melilotus*). Local species observed to be grazed include: caper

Table 5.1 Alphabetical List of the Flora of Tell Hesban and Area (number refers to page where species description is given)

Achillea santolina	85	*Demostachya bipinnata*	90	*Opuntia ficus-indica*	83
Agrostis sp.	89	*Dianthus strictus*	84	*Peganum harmala*	97
Alhagi maurorum	92			*Phalaris minor*	90
Amaranthus graecizans	82	*Echinops* sp.	86	*Phlomis* sp.	91
Amaranthus retroflexus	82	*Echium* sp.	83	*Phragmites communis*	90
Anchusa strigosa	83	*Erodium* sp.	89	*Pinus halepensis*	94
Artemisia herba-albe	85	*Erucaria* sp.	87	*Pistacia* sp.	94
Asparagus acutifolius	92	*Eryngium creticum*	97	*Polygonum equisetiforme*	94
Atriplex sp.	84	*Erysimum crassipes*	87	*Poterium spinosum*	95
Avena barbata or *A. sterilis*	89	*Eucalyptus* sp.	93	*Prunus amygdalus*	95
		Euphorbia sp.	89	*Pterocephalus pulverulentus*	88
Ballota undulata	91			*Reseda lutea*	95
Brassica sp.	87	*Ficus carica*	93	*Reseda* sp.	
Bromus sp.	89	*Ficus sycomorus*	93	*Rubus sanguineus* (=*R.sanctus*)	96
				Rumex roseus	94
Capparis spinosa	83	*Glaucium* sp.	94		
Casuarina stricta	84			*Salvia* sp.	91
Centaurea hyalolepis	85	*Heliotropium europaeum*	83	*Schinus molle*	82
Centaurea iberica	85	*Hordeum* sp.	90	*Scrophularia xanthoglossa*	96
Centaurea sp.	85	*Hypericum triquetrifolium*	91	*Silybum marianum*	86
Chenopodium album	84			*Solanum nigrum*	96
Chichorium intybus	85	*Lolium* sp.	90	*Sonchus oleraceus*	86
Chrozophora plicata	88	*Lycium* sp.	96	*Sorghum* sp.	90
Chrozophora tinctoria	89			*Stipa capensis*	90
Cirsium acarna	86	*Malva* sp.	93		
Cirsium syriacum	86	*Marrubium vulgare*	91	*Tamarix* sp.	96
Citrullus sp.	88	*Medicago* sp.	92	*Thymus captiatus*	91
Convolvulus arvensis	87	*Melilotus* sp.	92	*Tribulus terrestris*	97
Convolvulus dorycinum	87	*Mentha microphylla* or		*Trigonella* sp.	92
Crataegus sp.	95	*M. incana*	91	*Typha angustifolia*	96
Crepis aspera	86				
Crepis sp.	86	*Nerium oleander*	82	*Verbascum* sp.	96
Cupressus sempervirens	88	*Noaea mucronata*	84		
Cynodon dactylon	89			*Xanthium spinosium*	87
Cyperus longus	88	*Olea europaea*	93		
		Ononis natrix	92	*Zilla spinosa*	87
Delphinium sp.	95	*Onopordum* sp.	86	*Zizyphus* sp.	95

(*Capparis*), saltwort (*Atriplex*), common goosefoot (*Chenopodium*), lamb's quarters (*Chenopodium*), storksbill (*Erodium*), Bermuda grass (*Cynodon*), wild barley (*Hordeum*), wild rye grass (*Lolium*), canary grass (*Phalaris*), wild sorghum (*Sorghum*), feather grass (*Stipa*), and Medick (*Medicago*). Two species that are poor fodder, but which are locally grazed, are lovegrass (*Demostachya*) and camel thorn (*Alhagi*). One local forage species, common chicory (*Chichorium*), is reportedly very healthy for cattle.

Species Descriptions

AMARANTHACEAE

Amaranth or Cockscomb Family

Amaranthus graecizans L.
Tumbleweed
fis el kilaab, shagaret es-santeen
 This plant is found lying on the ground. It has tiny, purplish plume-like flowers which are clustered along the furrowed stem and surrounded by long, narrow leaves.
 Habitat. Waste places and roadsides in sandy soil.
 Season. Annual.

Amaranthus retroflexus L.
Pigweed, Hairy Amaranth
 An erect, slightly branched plant characterized by a terminal plume of numerous tiny, pale green flowers. The furrowed stem is covered with many short rough hairs. The broad, lance-shaped leaves are pale green in color.
 Habitat. One of the most common Amaranths, a weed of cultivation introduced from America. Found everywhere in waste places and tracksides. This plant belongs to the Borealo-Tropical group of segetal and ruderal species.
 Season. Annual; July-Sept.
 In the New World, the plants were cultivated by the Indians for their seeds which were ground into a flour. The leafy parts of the young plant may be eaten as a pot herb.

ANACARDIACEAE

Cashew Family

Schinus molle L.
Pepper Tree
shajarat ul-felfel
 This broad tree has a knotted, sometimes twisted, trunk with rugged bark. Its branches hang down like those of a willow tree. The leaves are compound with long and slender leaflets. The flowers are greenish yellow, forming small berry-like fruits in pendulous cluster, which have the size and taste of peppercorns. The fruit remains on the tree all winter.
 Habitat. Pepper trees are very tolerant of sun and periods of drought, making them most suitable for the Jordanian climate. In the Tell Hesban project area, they are found overhanging wadi beds where the streams are constant. They are also planted as shade trees around dwellings.
 Season. Deciduous; July-Aug.
 Originally a native to Peru, this tree is now a common ornamental in the Mediterranean area. It is a host to black scale which is detrimental to citrus trees. It is a fast growing tree which is self-seeding, giving it a tendency to become a weed. The leaves are rich in a volatile oil and when crushed, emit a distinctively peppery aroma, thus giving it its name. In the New World the Indians make beverages from its twigs and berries. The stem produces a gum-like substance called American Mastic which acts as a purgative. As an economic plant, it is utilized for its gum, as a medicine, and as a food for human consumption.

APOCYNACEAE

Dogbane Family

Nerium oleander L.
Oleander
dilfah
 This tall, erect woody shrub has many branches and leaves which are narrow, grey-green, and stiff and leathery. The fragrant white or pink 5-petaled flowers bloom continuously from the spring to the fall. The leaves exude a milky sap when broken.
 Habitat. Oleander is a dominant species of hydrophytic plants found on the banks of active stream beds. It was especially noted in the Wadi Hesban where it demarcates the course of the stream. It may also be found in gravelly places and damp ravines and can survive in almost any type of soil as long as the water table is near the surface.

Season. Evergreen; April-Oct.

Oleander is one of the most popular ornamental plants in the Middle East. All its parts are rich in highly poisonous alkaloids. It is known as "horse killer" in India and is used as a funeral plant in Christian and Hindu regions. The juice is used as an insecticide. When used as a medicinal plant, an active cardiac glucoside can be prepared from its sap. As a biblical plant it is possibly the "rose by the brook" mentioned in Ecclesiastes.

BORAGINACEAE

Borage Family
Anchusa Strigosa Labill.
Alkanet, Prickly anchusa
hamham

The stems of this plant are covered with stiff, prickly hairs. The plant itself is covered with rough bristles. The flowers are small, trumpet-like, and pale blue to violet to almost white in color. The seed pods are white and angular.

Habitat. One of the predominant species of the *batha-garigue* association, it is found along roadsides and on the periphery of cultivated fields. It is an Irano-Turanian element penetrating into the Mediterranean region.

Season. Perennial herb; April-June.

Echium sp. L.
Vipers bugloss
kahla

This is a hairy plant with a speckled stem covered with stiff rough hairs and having funnel-shaped flowers.

Habitat. Fields and stony ground, dry places.
Season. Perennial herb; April-July.

Heliotropium europaem L.
Heliotrope, European turnsole
sakran

Turnsole is a hairy plant with grayish, elliptical leaves, and white or lilac flowers similar to forget-me-nots.

Habitat. Cultivated ground, tracksides, waste ground and rocks, fields, and roadsides. A segetal-rudral biregional species of the Mediterrano-Irano-Turanian group.

Season. Annual; May-Sept.

CACTACEAE

Cactus Family

Opuntia ficus-indica (L.) Mill.
Prickly Pear Cactus
sabbayr

A succulent cactus plant with round, thick, racket-like fleshy leaves covered with sharp spines. The flowers are solitary, yellow, and showy.

Habitat. Arid and rocky places.
Season. April-July.

A very common plant, the cactus is a species introduced from the New World at the time of Columbus. It is used as an ornamental and as a fence around dwellings and kitchen gardens to keep out livestock. The fruits, which may be peeled and eaten, are about the size of a chicken egg. (They are sold in the market in Amman.) This plant is also planted along the roadside in the desert as a wind break and a barrier to keep the blowing sand off the roads. It propagates easily by planting any part in the soil.

CAPPARIDEACEAE

Caper Family

Capparis spinosa L.
Caper
ul asaf, asuf

This caper is a straggly, woody shrub with delicate, solitary, pink-white flowers having long dominant stamens. The grey-green, oval leaves are fleshy with curved spines at the base of each leaf. The branches are pendulant and hang from stone walls or cliffs.

Habitat. Semidesert, mountainous regions, rocky ground, steep cliffs. Found hanging from the rocky sides of dried-up stream beds in the Tell Ḥesbân project area, as well as on the tell itself on stone walls and rocks. A tropical genus of otherwise Sudanian origin.

Season. A summer perennial whose aerial shoots develop in summer; May-July.

The leaves and flower buds are eaten by the goats who are able to find it in precarious places. The flower buds are cooked and pickled to make capers. The small fruit, which is collected and eaten in the autumn, bursts when ripe, revealing seeds imbedded in jelly. Seeds of this plant have been found in quantity at Tell ea-Sawwan in Iraq with a date of about 5800 B.C. and are found also at the site of Beidha in Jordan. They act as a stimulus for the appetite.

CARYOPHYLLACE

Pink Family

Dianthus strictus (Banlis.)
kurunful

The commonest pinks found in the Eastern Mediterranean area, this plant has tiny toothed, single, pink flowers dotted with red spots. It has small narrow, lance-shaped leaves which are grass-like in appearance.

Habitat. Dry, rocky places. It was seen here and there among the ruins on the acropolis of Tell Ḥesbân but generally was not a common plant.

Season. Herbaceous perennial; May-June.

CASUARINACEAE

Australian Oak
Casuarina stricta
ait

This tall, sturdy evergreen with scale-like leaves looks like a tamarisk from a distance. Its grey-green branches look like thick pine needles, but are actually segmented as they have tiny leaves. Its small spiny cones are seen everywhere scattered along the road.

Habitat. Well-adapted to dry, warm climates and shallow, dry soils. It is a species introduced to the Near East from Australia.

This tree was seen being used as a fence or windbreak around olive groves and vineyards in the villages from Hesban to Madaba, as well as being used as a house screen along the roadside. It is also a good timber tree, but is probably not used as such in Jordan because trees are so scarce they are retained for their other uses.

CHENOPODIACEAE

Atriplex sp. L.
Orach, Saltwort, Purslane
qataf

This dwarf shrub has flat leaves which are pale on the underside.

Habitat. A halophytic shrub which may indicate soil salinity.

Season. Germinates in the spring.

The leaves are eaten by animals. *Atriplex* may be cooked and eaten like spinach, and could be the mallows referred to in Job 30.

Chenopodium album L.
Common Goosefoot, Lamb's Quarters
fiss-ul-kilaab

Goosefoot is an erect herb with mealy white leaves and a spike-like flower.

Habitat. Common in waste places. It is part of the Borealo-Tropical group, a ruderal, pluri-regional species. *Chenopodium* is commonly found as a weed on cultivated land, especially as an impurity in clover and cereal crops.

Season. Annual; May-Oct.

The leaves and stems of this plant are eaten by animals as well as used as greens by humans. Each plant produces as many as 3,000 seeds with high protein and fat content. It is also rich in carbohydrates and may be used as a cereal substitute. The ground seed has been known to be used in place of flour in years of famine. The greens may be boiled and eaten. Remnants have been found in the Early Bronze Age lake dweller's sites in Switzerland.

Noaea mucronata (Forssk.) Asch. et Schweinf.
Thorny saltwort
shawk-ul-hanash

Like *Poterium*, this woody shrub has flowers at the base of the thorny twigs, losing its large winter leaves in the summer, and replacing them with smaller summer leaves in an effort to reduce transpiration.

Habitat. This plant is one of the main components of dwarf shrub steppes. It is widely distributed throughout steppes and desert. At Tell Hesban, it is found in the rocky places near the tell and in the hills east of the tell. It is most often associated with *Poterium* and *Ononis*.

Season. July-Oct.

COMPOSITAE

Daisy Family

Achillea santolina L.
Milfoil, Yarrow
bishreen

This is an erect, green, hairy plant whose leaves have tiny segments. The yellow flower heads have short rays.

Habitat. A representative of the *Mauritanian Steppe* element in Palestine and part of the *Artemisia herba-alba* association of the Irano-Turanian territory. Characteristic of loose soil, but in the Hesban area, found on dry hills and sandy places.

Season. Perennial; April-June.

A fragrant herb, yarrow is used medically as a blood thinner and to induce bleeding.

Artemisia herba-alba L.
Sagebrush
sheeh

Sagebrush is a small, woody, grey-green shrub with wooly textured stems and small, narrow leaves. It has a shallow root system of fibrous, vertical branches. A pleasing odor is given off when the plant is crushed.

Habitat. The dominant and typical species of the Irano-Turanian territory, it is found on rocky hillsides in limestone and calcareous soils, loess, and sand. It was not seen in great abundance in the Hesban area.

Season. Perennial. Flowers and fruits in early winter, partially dies back in summer.

Artemisia has medicinal uses.

Centaurea sp. L.
Centaury
shawk ud-dardar

This thistle has reddish-purple flowers and conspicuous yellow spines surrounding the bottom of the flower. This particular specimen was difficult to identify on a species level because of a tendency towards polymorphism and regional variations.

Habitat. A common Irano-Turanian plant found on tracksides, fields, and uncultivated ground.

Season. Biennial; May-July.

The young stems and leaves of this plant are edible.

Centaurea hyalolepis Boiss.
Star thistle
shawk

This is a straw colored spiny thistle with sulphur colored flowers surrounded by four or five sharp spiny projections.

Habitat. Waste places. In the Hesban area, commonly seen along tracksides and the edges of agricultural fields.

Season. April.

Centaurea iberica Trev. ex Spreng—hybrid with *C. hyalolepis*
Thistle
shawk

This thistle has yellow flowers and physical characteristics similar to the previous specimen.

Habitat. In the Hesban project area, found in the narrow strip between the road edge and the parallel ditches behind.

Medicinally, this plant is used as a cure of ague and jaundice. It is also a nutritious vegetable.

Chichorium intybus L.
Common Chicory
shikuriyyah, handab

Chicory is a lovely, bright blue dandelion-like flower with fringed petals. This plant can reach the height of four feet or more. The flowers are short-lived, blossoming in the morning and wilting by evening. The lance-shaped leaves have hairy undersides.

Habitat: Road and tracksides, waste places and uncultivated ground; often found on chalk or limestone. Originally native to Britain.

Season. Perennial; May-Sept.

The roots of chicory can be dried and roasted as a coffee substitute or, as in French coffee, as an additive to give a bitter taste. The roots are also sometimes cooked as a vegetable. The leaves may be used as a salad plant. It is considered valuable in pasture for the health of the cattle. It is favored by honey bees as a source of nectar. As a medicinal plant, it is supposed to stimulate the appetite and is also used as a tonic or diuretic. It has also been regarded as an aphrodisiac.

Cirsium acarna (L.) Moench
Thistle, Yellow Cnicus
shawk ul-far

This thistle has purple flowers densely clustered and enclosed by very spiny upper leaves. The leathery leaves are linear with long yellow spines at the end and at regular intervals on the sides.

Habitat. Fields and stony places, waste places.

Season. Perennial; July-Aug. *Cirsium* spp. are biseasonal perennials which develop their vegetative leafy shoots in winter and flowering shoots in summer.

Cirsium syriacum Gaertn. (*Notobasis syriaca* [L.] Cass.)
shawk ul-hanash

This is a stout thistle with naked branches. The leaves below the head have long red or white spines from the midrib and side veins. They may also have white mottling. The long flower heads are egg-shaped.

Habitat. Tracksides, poor uncultivated ground, and rough fields.

Crepis sp. L.
Hawkweed

The specimen collected is a flower head gone to seed—soft, white hairs (pappus) which float on the air. The flower probably was yellow, similar to a dandelion.

Habitat. None of the *Crepis* species found in the Hesban area are native to the area. Found in disturbed but uncultivated areas, in rocks around olive groves.

Season. Annual.

Crepis aspera L.
Hawksbeard

This specimen is a tall member of the *Crepis* genus with small, yellow, bristly, dandelion-like flowers and toothed leaves. The stems and branches are very rough with rigid, prickly bristles.

Habitat. Weedy places, roadsides.
Season. Annual; July-Aug.

Echinops sp. L.
Globe Thistle
shawk ul-jimal

This is a showy thistle with a large, light blue, globe-like head composed of many tiny flowers. The leaves and stems are spiny.

Habitat. An Irano-Turanian derivative with a vertical, thick tap root which goes down 1-2 m to acquire water in the dry season. It was seen in the tell, and along tracks and fences in the village of Hesban. Dry, rocky places, uncultivated ground.

Season. Perennial; July-Sept.

Onopordum sp. L.
Cotton Thistle
shawk

These are tall, spiny herbs with violet flowers with large heads. They also have a deep unbranched vertical tap root.

Habitat. An Irano-Turanian derivative. Fields and waste places.

Silybum marianum (L.) Gaertn.
Holy Thistle, Milk Thistle
khurfaysh ul-jima

This thistle has a large, solitary, purple flower surrounded by sharp pointed bracts. The leaves and bracts have spiny points at the tip. The leaves are mottled with white, and are dark and shiny with deeply cut triangular lobes. The furrowed stem has many small thorns.

Habitat. Rocky, uncultivated ground, tracksides—occupies the narrow strip between the road edge and the parallel ditches behind. Formerly a native of Britain where cultivated. A biregional species from the Mediterrano-Irano-Turanian group.

Season. Perennial; April-Aug.

The young leaves of this plant are used as salad components by the Arabs. The flower part may be cooked and eaten like an artichoke.

Sonchus oleraceus L.
Sow Thistle
hawa

This herb has yellow, dandelion-like florets.

Habitat. Found on loose heaps of debris and in the dust near human dwellings.

Season. Annual. Summer weed.

This plant is mentioned as an esculent by Dioscorides. The young leaves may be put into salads and eaten like dandelion greens.

Xanthium spinosum L.
Spiny Clotbur
shobbei

This plant has small heads of green flowers and spiny burred fruits. The spines are long, yellow, and needle-like. The green, hairy fruit is covered with bristling, narrow hooked, yellow spines.

Habitat. Waste places, tracksides, banks, and hedges. Introduced from South America through Portugal. The hooks on the fruits attach onto the coats of sheep and goats, and are therefore dispersed by animals.

Season. Annual; July-Sept.

CONVOLVULACEAE

Bindweed Family

Convolvulus arvensis L.
Field Bindweed
meddaad, olleiq

Bindweed is seen as a vine with arrow-shaped leaves which trails along the ground. The flower is a white trumpet with pink to red stripes. The trumpet closes in the dark and in bad weather. This common weed propagates itself by rhizomes which can penetrate the earth to 2 m in depth.

Habitat. Found on the edges of cultivated fields, along tracksides, and waste places. Sometimes even seen climbing up plants in the field.

Season. Perennial; June-Sept.

Bindweed is a popular source of nectar for honey bees. Used medicinally, it is a powerful purgative.

Convolvulus dorycinum L.
Morning Glory

This is a tall herb with very few leaves on the lower parts. The trumpet-like flower is light pink.

Habitat. Nonirrigated cultivated fields; also noted on the trackside in the Wadi Hesban.

Season. Perennial; July-Sept.

CRUCIFERAE

Mustard Family

Brassica sp. L.
Mustard

This is a tall weed with small, yellow flowers at the top of the plant. There are many small, slender, pod-like fruits branching off the stem beneath the flowers.

Habitat. Often found associated with nonirrigated crops, and sandy and clay soils.

Season. Annual.

Erucaria sp. Gaertn.

This weed has purple flowers in racemes without leaves. The seedlings of the plant often occur among summer irrigated crops.

Habitat. Found associated with both irrigated and nonirrigated crops.

Season. Winter annual.

Erysimum crassipes

Habitat. One of the few Irano-Anatolian species in Jordan, usually confined to higher altitudes of Transjordan.

Zilla spinosa (Turra) *Prantl*
zilla

This is a low, stiff branched plant with spines that grow at right angles to the stem. The leaves are small and fleshy, and in many cases dried out or nonexistent. The four, petaled flowers are pale pink to dark pink.

Habitat. Seen predominantly in harvested wheat fields just west of Jalul. This plant seems to be avoided by grazing animals. It was observed in soil that is nonirrigated, shallow, and quite rocky. This is one of the most important shrubs in the Saharo-Sindian flora group and one of the few represented in the flora of the Tell Hesban region. Ordinarily it would be confined to interior sand dunes and sandy plains.

CUCRBITACEAE

Cucumber Family

Citrullus sp. (probable *C. colocynthis* [L] Schrad.)
Bitter Gourd
handal

This plant was found sprawling on the periphery of cultivated areas. It has hairy stems and leaves, making it relatively unpalatable to grazing animals. The fruits are like small, green golf balls.

Habitat. A desert plant which has penetrated the alluvial plains. Its center of origin is believed to be South Africa.

The bitter fruits have a purgative effect and therefore may be used medicinally for this purpose. Animals also avoid eating the plant for this reason.

CUPRESSACEAE

Cypress Family

Cupressus sempervirens L.
Mediterranean Cypress

This tall, evergreen tree grows vertically in a columnar fashion or horizontally in the shape of a pyramid. The domesticated specimens tend to grow in the vertical manner. The feathery scale-like leaves secrete a resin which has medicinal uses. The bark of the older trees is markedly furrowed and fibrous.

Habitat. This tree has a long life expectancy. It is drought resistant and likes warm climates. It is adaptable to any sort of soil. In the Tell Ḥesbân project area, it is found in locations similar to those of *Casuarina stricta*. As an ornamental, it is planted around houses and yards.

Season. Perennial; April.

The wood of the cypress is resistant to rot, and was used by the ancients to manufacture storage boxes. Its hardness and durability made it especially popular for shipbuilding and house construction. The overuse of this wood in ancient times makes any wild stands of the cypress a rarity. It was referred to as gopher wood in the Bible.

CYPERACEAE

Sedge Family

Cyperus longus L.
Galingale
se'ed

This is a tall grass-like plant with sizeable spikelets on the ends. The plumes are reddish-brown in color and arranged in an umbrella pattern. The leaves are along the stem.

Habitat. Ditches near water or marshes. Frequent in wet areas. Noted on the banks of flowing streams, as at Wadi Hesban. Part of the *Phragmites-Typha* association commonly found at springs and on banks of both permanent and ephemeral streams.

Season. Perennial; March-Aug.

DIPSACACEAE

Scabious Family

Pterocephalus pulverulentus
lisseq

This is a herbaceous plant with a dense flower head.

Habitat. Found in the immediate environs of Tell Hesban on dry, shallow, rocky soil. This plant and *Erysimum crassipes* are the only Irano-Anatolian species found here.

EUPHORBIACEAE

Spruge Family

Chrozophora plicata (Vahl) A. Juss. ex Spreng.
Plaited Leaved Croton
koddah

Croton is a wooly plant with bunches of tiny, green flowers. The leaves are triangular and heart-shaped. They are grey-green with a lighter underside and are velvety in texture, covered with numerous small hairs giving the leaves and stems a hairy furry appearance.

Habitat. Found consistently as a weed in tomato, tobacco, and okra fields; waste areas. Found also in seasonally watered depressions in sandy fields.

Season. Annual; April-Nov.

Chrozophora tinctoria (L.) A. Juss. ex Spreng
Turnsole
faqqoos el-homaar

This is a plant with grey-green, furry leaves which are ovate or rhomboid in shape. The flowers appear in yellow clusters. The fruit is small, blue, and berry-like.

Habitat. Found in cultivated ground or waste places, especially in *terra rossa* or rendzina soils.

Season. Annual; June-Oct.

The small, blue berries are a source of "turnsole" dye. The seeds may be used medicinally as emetics.

Euphorbia sp. L.
Spurge
halib ul bum

This euphorbia is a woody erect shrub with bluish-green leaves. The small, green flowers are in ray-like heads surrounded by cup-shaped bracts or umbrella-shaped "umbels." The narrow lance-shaped leaves are spirally arranged round the stem. The shrub exudes a bitter, milky juice when broken, making it unpalatable as a forage plant.

Habitat. Seen often on rocky, open terraced fields which have presumably been cleared for cultivation. One of the richest genera represented in Jordan.

The milky juice is poisonous and has a violent purgative action. It is used medically in small doses. It is, therefore, one of those plants avoided by animals and thus survives where more palatable species would be consumed.

GERANIACEAE

Geranium Family

Erodium sp.
Storksbill
qarni morghaat

These plants have lobed leaves with pinnate veins. The flowers are arranged in umbels. The fruit has a long beak, thus the name.

Habitat. Road edges and parallel ditches.

Erodium is cultivated as a forage plant in some areas.

GRAMINEAE

Grass Family

Agrostis sp. L.
Bent Grass

This is a tall grass with a rich panicle of tiny, single-flowered spikelets. The flowers are either greenish or purple, to brown in color.

Habitat. Roadsides or meadows, in moist places, and ditches. Found growing in the banks of flowing streams of Wadi Hesban.

Season. Perennial.

Avena barbata Brot. or *A. sterilis* L.
Wild Oat
bohma

This grass has awns consisting of two to four flowered spikelets.

Habitat. A weed found everywhere in the Hesban area—in fields, fallow ground, olive groves, vineyards, tracksides, and roadsides.

Season. Annual; April-June.

Avena ia a pasture plant which is palatable and abundant in the rainy season.

Bromus sp. L.
Brome Grass
khafur

Brome is a low, sparsely branched grass with an ovate, erect head which has many long awns, giving the head a feathery appearance. The leaves are inconspicuous.

Habitat. Dry, waste places, uncultivated ground, edges of fields, walls, and fences.

Season. Annual.

Brome is a pasture plant which is abundant in the rainy season.

Cynodon dactylon (L.) Pers.
Bermuda Grass
meddaad, negil

This is low, grayish-green grass which creeps over the surface of the ground sometimes forming dense mats. The short stems bear a group of four to five purple or green finger-like spikelets. This grass propagates itself by rhizomes and thus is difficult to destroy.

Habitat. A very common weed found on uncultivated ground, dry places, and tracksides. Indicative of a high ground water table.

Season. Perennial; May-Aug.

Because of its ability to survive intense grazing, this grass is an especially important forage plant.

Demostachya bipinnata (L.) Stapf.
Lovegrass
halfa

Halfa is a very tall, rigid grass with spikelike panicle of white to pink flowers and long leaves. It has an extensive rhizome underground root matrix which enables it to reproduce vegetatively underground and cover an extensive area.

Habitat. Found on the banks of flowing streams associated with typha, oleander, tamarisk, and other hydrophilic plants. Also found in waste places.

Season. Perennial; May-Dec.

It may be grazed if more palatable plants are not available.

Hordeum sp. L.
Wild Barley
bahma, sha'eeriyyah

Hordeum is a tall, hardy grass with a bristly, dense spike at the end of the stem, bearing many seeds in the head. It resembles a spike of cultivated barley, but is not as full. The leaves are flat and near the base of the stem.

Habitat. Waste places, a possible escapee from crop plants.

Season. Annual; June-July.

Wild Barley is a forage plant which is grazed in the rainy season.

Lolium sp. L.
Wild Rye Grass
sammh, siwan

This is a medium sized grass with flat leaves. The seeds appear in clusters on the spike opposite to each other up to the tip of the spike.

Habitat. A common weed of grain crops, found in cultivated areas.

Season. Annual; April-June.

The seeds of *Lolium temulentum* are referred to in the Bible as "tares". They are said to be poisonous, however, the nonpoisonous species are popular forage plants.

Phalaris minor Retz.
Canary Grass
sha'eer el-faar, 'ain el-gott

Phalaris is a grass of medium height with a dense, compact head.

Habitat. Around dwellings.

Season. Annual.

Canary grass is used for grazing in the rainy season when it is especially abundant.

Phragmites communis (L.) Trin. (*P. australis*)
Common Reed
aghanim, ghab, bus

This reed is a grass up to ten feet tall with long, silky, plume-like flowers. The flat, grey-green leaves are two inches wide. The stiff, nonwoody stems are hollow. The plants propagates from rhizomes.

Habitat. On the borders of ponds or swamps. In the Hesban area seen at the edges of flowing streams, growing in clumps.

Season. Perennial; April-Sept.

The dried stems may be used for mats and thatching.

Sorghum sp. Moench
Wild Sorghum

This is a broad-leafed grass with loosely branched erect heads which are purple in tint.

Habitat. In dry fields and along roadsides.

Season. Annual.

Some species of sorghum are cultivated for fodder. It is a popular grazing grass which may be an escapee from cultivation.

Stipa capensis Thunb.
Feather grass
sabat

This is a tufted grass with convolute or flat leaves.

Habitat. Dry, rocky places, cracks in rocks where moisture accumulates. A desert and steppe grass.

Season. Annual. March-June.

Stipa is popular as a forage grass.

HYPERICACEAE

Saint John's wort Family

Hypericum triquetrifolium Turra (*H. Crispum* L.)
St. John's wort

This is a low heather-like shrub with small, yellow 5-petaled flowers and lance-shaped leaves. The flowers are arranged in a candlelabra-like formation on the shrub. The plant gives off a spicy, resinous aroma when crushed.

Habitat. A *maquis* plant which favors rocky places. Seen in olive groves, vineyards, and waste ground.

Season. Perennial; late summer, June-Aug.

Medicinally, it is used by the Arabs in making a kind of tea, the steam of which is supposedly good for respiratory ailments.

LABIATAE

Mint Family

Ballota undulata (Sieb. ex Fres. Benth.)
Wavy Ballota
ghassa

This is a gray-green wooly plant with round, trumpet-shaped velvety leaves. The single stems that are derived from a common root base stand erect. The flowers are pale pink to white in color, and are arranged in globe-like clusters along the stem.

Habitat. Very commonly seen on the tell and in crevices in rocks along the tracksides. Along with burnet and thyme, *ballota* is one of the leading plants of the Mediterranean *batha* community.

Season. Perennial; spring.

Marrubium vulgare L.
White Horehound
robeia

This wooly, gray-white, erect plant sometimes grows to 60 cm. It has clusters of small, white flowers surrounded by wooly crinkled leaves.

Habitat. A hardy plant, well-adapted to full sun and intense heat, it is found on dry slopes, paths, and waste ground. A biregional species of the Mediterranean and Irano-Turanian groups.

Season. Perennial; April-Sept.

This plant, which is originally native to Britain, can be made into a tea which may be used as a stimulant and to reduce fever. It is also used as a remedy for coughs.

Mentha microphylla Koch. or *M. incana* L.
Mint
nahnah

This tall erect plant has pale blue lavender flowers densely arranged on a narrow spike. The broad serrate leaves are arranged in pairs at the base of the spikes. The plant gives off a significantly minty smell when crushed.

Habitat. Found alongside the flowing streams of Wadi Hesban, along with the grasses and oleander; generally found in wet places.

Season. Perennial.

The leaves of this plant are used to give tea mint flavor.

Phlomis sp. L.
Phlomis
zeheira

This hardy yellow-green plant has dense, wooly stems and elongated heart-shaped leaves.

Habitat. Seen on the edges of cultivated fields and along tracksides in calcareous soils. It is a xerophytic steppe plant.

Season. Perennial; May-July.

Salvia sp.
Sage
kharnah

Salvia is a woody shrub with gray-green aromatic leaves and pale blue to white flowers.

Habitat. A *maquis* and steppe plant, it was seen in rocky areas and dry places, usually on limestone. It likes sun.

Season. Perennial; March-July.

The aromatic leaves may be burned for fumigation purposes or may be used in making tea or as a seasoning. It is a popular nectar plant for bees.

(Corido) Thymus capituatus (L.) Hoffmgg. et Link
Headed Savory
za'tar

This dense shrub with thick, spiny, woody branches gives off a pleasant aromatic smell when crushed. The short, stiff, linear leaves are summer leaves. They replace larger winter leaves which are shed in warm weather to reduce transpiration. The blue-purple flowers are found at the end of the stems.

Habitat. In the project area, this plant was seen on the hill opposite Tell Hesban, as well as in the vertical banks of the dried-out stream beds. It prefers sunny hills and rocky places as well as limestone and clay soils. It is a characteristic species of the seral community preceding a forest and maquis climax.

Season. Perennial; May-Sept.

This aromatic plant is often used on floors to give a pleasant odor to homes with dirt floors. It is also used as a seasoning. The essential oil derived from this plant is used in medicine and as a perfume base. It is also a good honey plant.

LEGUMINOSAE (PAPILIONACEAE)

Pea Family

Alhagi maurorum Medic.
Camel Thorn
shawk ul jimal, 'aqool

This is a low, woody shrub with pink flowers and small leathery leaves. Its spiny-tipped branches grow at many angles, somewhat reminiscent of burnet. It has tiny pod-like fruits. *Alhagi* propagates itself by creeping rhizomes, making it a plant difficult to destroy. It is a deep rooting plant which is able to seek out water during dry periods.

Habitat. Weed in waste places. In the Tell Hesban area, this plant was seen most often in cleared fields that have been repeatedly cultivated and heavily grazed over by animals. It is an indicator of an accumulation of salts in soil due to evaporation and improper drainage. Dominates Mediterranean cultivated alluvial soils.

Season. Perennial; June-Sept.

This plant was observed being eaten by sheep, goats, and camels, in spite of its spiny twigs. In some areas it is also dried and used as fuel. Boulos says that it is boiled in water by the Nubians to prepare an extract for the treatment of bilharzia and rheumatism.

Medicago sp.
Medick
nafal

The dentate leaflets of this plant are arranged in groups of three. Its distinctive spiral seeds are seen everywhere on the ground along tracksides, even when the plant itself is not evident. The seeds resemble those of *M. orbicularis*.

Habitat. Cultivated ground, olive orchards. Part of the *Poterium* community in the common *batha* association.

Season. Annual; March-June.

Since this plant is a member of the clover family, it is especially valuable for fixing nitrogen in the soils. It is also a popular forage plant and is a component of natural pasturage in the winter months.

Melilotus sp. L.
Sweet Clover, Melilot
handakuk

This is another nitrogen fixing plant with leaves consisting of three dentate leaflets. It has a smell like coumarin when crushed.

Habitat. Some species prefer chalky or gravelly soils, uncultivated dry places, fields, and tracksides.

Season. Annual or biennial; May-Aug.

Because this plant contains coumarin, it is sometimes used medicinally as a blood thinner. It is also a popular fodder plant. The pleasing aroma of its crushed leaves makes it useful as a deodorizing plant on house floors.

Ononis natrix L.
littein

This is a dense, woody shrub with spiny branches and large yellow pea-like flowers.

Habitat. A plant of the semisteppe and Irano-Turanian community. Commonly found on rocky slopes near the tell, mixed with the *Poterium* community. Also found on sandy and stony places and calcareous soils.

Season. Perennial; April-July.

Trigonella sp. L.

This plant is very similar to melilot. It has three foliate dentate leaflets and yellow pea-like flowers. It also smells like coumarin when crushed. It has small, sickle-shaped pods.

Habitat. Dry hills.
Season. Annual.

LILIACEAE

Lily Family

Asparagus acutifolius L.
Sharp-leaved Asparagus
shawk

The leaves of this plant are reduced to scales, giving it a feathery appearance. It is a shrub-like plant with woody stems and many branches. There are small spines in its lower leaves. It has a small, black, berry-like fruit and a fleshy root.

Habitat. Stony ground and dry places. Found mainly on limestone on tracks above the edges of the wadis.

Season. Perennial; Aug.-Dec.

The young shoots can be eaten although they may be a bit bitter and stringy.

MALVACEAE

Mallow Family

Malva sp.
Mallow

This is a herbaceous plant with broad palmate leaves and pink-purple flowers.

Habitat. Fallow land, tracksides, and cultivated ground.

Season. April-Aug.

When the leaves of this plant are boiled, a mucilaginous substance is produced. The leaves and young shoots are used as food. Okra ia a member of this genus.

MORACEAE

Mulberry Family

Ficus carica L.
Fig
teen

This tree is often cultivated for its fleshy fruit. It has large three to five lobed leathery leaves. The fruits are edible and about the size of golf balls.

Habitat. Near springs in garden spots or on dry hillsides in groves with almond trees. Likes full sun. Can grow in poor or shallow soil.

Season. Deciduous.

Although the fig grows wild, it is cultivated for its fruit, which bears two crops. The fruits may be used medicinally as a laxative or dried as a poultice for wounds and boils.

Ficus sycomorus L.
Mulberry Fig, Asses Fig
glomez, gemmeiz

This is a broad tree with large, heart-shaped leathery, olive green leaves and yellow bark. The fruit is small, about the size of a cherry.

Habitat. Can tolerate drought and heat, but prefers a moist location.

Season. Deciduous.

The wood of this tree is especially valued for its hardness and was used for the manufacture of sarcophagi in ancient Egypt. The leaves are used in the preparation of a diuretic, and the fruit may be eaten.

MYRTACEAE

Myrtle Family

Eucalyptus sp. (probably *E. rostrata*)
Eucalyptus, Gum Tree
kafour

This is a tall tree with long, spear-shaped leaves and white powder-puff flowers. The gray bark appears mottled and is easily peeled.

Habitat. A hardy plant which tolerates even salty soils. It is commonly cultivated and was introduced from Australia in the 1800s. Seen along stream beds in the Tell Hesban project area.

The flowers of this tree are a source of honey nectar.

OLEACEAE

Olea europaea L.
Olive
zeitun

These trees have curiously twisted trunks covered with smooth, gray bark which cracks with age. The leaves are lance-shaped with a leathery texture and are greenish-blue above and silvery on the underside. The fruits are black drupes.

Habitat. Will tolerate drought and likes sun. Can grow in shallow or poor soils. Found wild throughout Mediterranean countries. Possibly an escapee from cultivation.

Season. Evergreen.

A very important plant since ancient times, the olive is a primary source of oil for domestic use. The fruit is also pickled and eaten. The olive is a symbol of peace and is referred to often in the Bible. Ancient oil presses, carved from the local limestone, were seen abandoned in the hills above and surrounding Wadi Hesban.

PAPAVERACEAE

Poppy Family

Glaucium sp.
Poppy

This plant was found as a singular specimen. It was conspicuous with a bright yellow waxy flower with four fragile petals. A plant which lies close to the ground, it has large irregular blue-grey leaves and rough, hairy, short stems.

Habitat. Limestone hills and waste places. This specimen was seen on a hillside and trackside overlooking a watering place.

PINACEAE

Pine Family

Pinus halepensis Mill.
Aleppo Pine

This medium sized tree has green paired needles for leaves. The tree is generally umbrella shaped and not particularly tall. The trunk is twisted, and the bark is ash grey when young, turning to brown. The fruits are ovate cones.

Habitat. Likes highly calcareous soils and is found wild in limestone hills. It is a long-living tree which has become important in reforestation projects. Drought resistant.

Season. Perennial; March-May.

Because of its strong durable wood, this tree was important in ancient times as a timber tree. It was used in shipbuilding and house building. It may be tapped as a source for resin and pitch. Tannin for the dressing of skins and hides is also derived from its bark. Its use as an important fuel source accounts for its gradual disappearance from the landscape.

PISTACIACEAE

Pistachio Family

Pistacia sp. L.
Pistachio

This tall tree with shiny, green, leathery leaves bears a small nut-like fruit. It may attain great age and can grow as tall as 20 m.

Habitat. Dry climates; the same type of environment as olive trees. Where found in the wild, the pistachio is often associated with Irano-Turanian dwarf shrubs and species. It is a fast growing tree native to the Middle East. In the Tell Hesban project area, it was seen principally in a stand of forest north of the tell near *Naur* in association with Aleppo pines and species of oak.

Season. Deciduous.

Some species of pistachio are cultivated for their edible fruit.

POLYGONACEAE

Dock Family

Polygonum equisetiforme Sibth. et Sm.
Knotweed
gordaab

This is an erect or trailing herb with a woody root. Its tiny, dark pink flowers are located in the axil of the leaf along the stem. The leaves are small and linear.

Habitat. In the Tell Hesban area, this plant is found as a weed in sandy soil along the road and trackside, and in the ditch beside the main road.

Season. Perennial; July-Oct.

This herb is rich in tannin and silicic acid, making it useful as a poultice. In ancient times, it was used medicinally as a remedy for diarrhea and dysentery as well as for circulatory illnesses.

Rumex roseus L.
Dock
hommaad

Dock has fleshy, oval leaves. The flowers are arranged in a terminal raceme without leaves.

Habitat. Rumex is a member of the desert and *hammada* plant community. It is found on rocky, dry soils.

Season. Annual.

The leaves and stems of dock are edible and may be used as greens.

RANUNCULACEAE

Buttercup Family

Delphinum sp. L.
Larkspur

This delicate, grass-like plant has a violet-blue flower with a spur. The erect, leafless stem is covered with fine hairs.

Habitat. This single specimen was seen on the goat track beside the dried-up stream bed in the rocky hills below Ain Sumiya. It favors dry, rocky places.

Season. Annual; May-Aug.

This plant is very poisonous and should be avoided by grazing animals, although goats are known to be able to tolerate it. Medicinally, it has been used as a pain reliever. Economically, it has been used as an insecticide.

RHAMNACEAE

Buckthorn Family

Zizyphus sp. (either *Z. spina-christi* or *Z. lotus*)
jujube
nabq, gabaat

A small broad tree with pale green, oval leaves and small, greenish white flowers. The fruits are small and berry-like. Small curved thorns are arranged in pairs at the base of the leaves.

Habitat. This tree was seen as a solitary tree in the middle of cultivated fields. These fields were located in the highlands near Ain Musa and near the Naur-Jerusalem road.

Season. Deciduous; May-July.

The tree is considered a sacred tree and is often placed in the midst of a cultivated field to provide shade for the farmer as well as luck for a prosperous harvest. On one occasion, a tree had ribbons and belongings tied to its branches, possibly as offerings in prayer. The small fruits are edible.

RESEDACEAE

Mignonette Family

Reseda lutea L.
Wild Mignonette
weybe

This herb has tiny pale green flowers in a spike-like raceme. The pale green leaves are deeply lobed.

Habitat. Fallow ground, stony places, banks and tracks, especially on limestone soils.

Season. Biennial; April-July.

ROSACEAE

Rose Family

Crataegus sp.
Hawthorne

This thorny, woody shrub has small, leathery leaves and large, red, berry-like fruits.

Habitat. Found all over the limestone hills in the same association with *Poterium* and *Ononis*.

Season. April-May.

The fruits of the hawthorne are edible.

Poterium spinosum L.
Thorny Burnet
billan

This low dense woody shrub has many branches which are positioned at right angles to each other. The branches end in double spines. The small red flowers are evident in the spring and turn to small red-brown berries which were seen everywhere on the tracks. Its large winter pinnate leaves drop and change to small summer leaves during the hot, dry season.

Habitat. Dry limestone hills, especially where forest and maquis have been cleared. This is the dominant plant of the area around Tell Ḥesbân. It is found virtually in all ditches except in the heavily watered areas where other species are able to compete effectively.

Season. Perennial; March-April.

This plant is commonly cut in the summer for fuel to be used particularly in the bake ovens and lime kilns. The plants are torn up by the roots and piled in large piles to dry. They also act as effective natural fencing for the bedouin animals at night. The plant is referred to in the Bible in Isaiah 34.

Prunus amygdalus Batsch
Almond
loz

Habitat. Growing both wild and cultivated, was found mixed in olive and fig groves near

Tell El Al. It is very tolerant of poor and shallow soils. Introduced from China into Europe in the 6th and 5th centuries B.C.
Season. Deciduous.
The fruits of this tree are edible nuts and may also be used as a source of oil. The Romans ate them before drinking to avoid intoxication.

Rubus sanguineus Frivaldszk. (=*R. sanctus* L.)
Blackberry
al lig
This green thorny plant has broad, rounded leaves with serrate edges. The leaves occur in groups of three. The small rose-like flowers are pink or white and produce a black, compound fruit. The woody stem has short straight or curved spines.
Habitat. Grows in thickets alongside streams and on river banks.
Season. Perennial; April-June.

SCROPHULARIACEAE

Figwort Family

Schrophularia xanthoglossa Boiss.
Figwort
barwek, qortom
This plant has distinctive arrow-shaped leaves and large yellow flowers.
Habitat. Cultivated ground, vineyards, walls, rocky, and stony places.
Season. Perennial.

Verbascum sp. L.
Mullein
kharma
This light yellow-green, wooly plant has small yellow flowers on a tall, erect spike coming from the center of the plant. Its fuzzy leaves have undulating edges.
Habitat. Found everywhere on uncultivated land growing out of the cracks in the limestone and along fences and walls. Dry soil.
Season. Biennial; May-Aug.
Since some species are poisonous, they are selectively *not* grazed by animals, so they have an opportunity to successfully compete for niches with other plants which have been cleared or grazed. The flowers were used as a source of yellow dye in ancient times.

SOLANACEAE

Nightshade Family

Lycium sp.
'awsaaq, hawshez
This thorny, dense woody shrub has small spatulate leaves. The fruit is a small, hard, red berry about the size of a pea. The flowers were not collected.
Habitat. Found in association with the woody shrubs on the limestone hillsides in the vicinity of Tell Hesban.

Solanum nigrum L.
Black Nightshade
Banduret deeb
This is an erect herb with ovate leaves and small, white, starry flowers. The fruit is a hard black berry less that 1 cm in diameter. The stem is hairy.
Habitat. Cultivated ground, waste places, and tracksides.
Season. Annual; Jan.-July.
The berries of this plant are especially poisonous. They contain the lethal alkaloid solanin. The leaves, however, may be boiled and eaten like spinach.

TAMARICACEAE

Tamarisk Family

Tamarix sp. L.
Tamarisk
hatab
This is a graceful tree or large shrub. The tiny blue-green, scale-like leaves give the tree a feathery appearance. The tamarisk has a vertical tap root which enables it to seek out water.
Habitat. Alongside streams and damp places.
Season. Deciduous; March-July.

TYPHACEAE

Reedmace Family

Typha angustifolia L.
Lesser reedmace, Cattail
halfa, bordi
This marsh plant often grows taller than a

person. It has long, narrow, ribbon-like leaves. The brown, furry spike at the end is the cattail.

Habitat. In the Hesban area, it was found in ditches and wet places in the periphery of active streams in the same communities with *Phragmites*.

Season. Perennial.

The young roots of this plant are edible.

UMBELLIFERAE

Carrot Family

Eryngium creticum Lam.
Blue Eryngo
kurs'anni

This spiny herb resembles members of the thistle family. It has head-like blue flowers surrounded by spiny projections. The leaves are mostly absent during the summer months. The entire plant is of a blue-green color.

Habitat. Seen virtually everywhere in dry, stony places, tracksides, and uncultivated ground.

Season. Perennial; May-Aug.

The roots of this plant are sometimes used against snake bites. The rosettes which last through the winter are sometimes eaten as a vegetable. The young leaves may be used as a pot herb.

ZYGOPHYLLACEAE

Caltrops Family

Peganum harmala L.
Common Peganum
harmal

The yellow-green trilobed seeds of this plant are surrounded by thin, leafy bracts. The leaves are long and thin and extend from the base of the branches on which the fruit is born.

Habitat. Roadsides and sandy places, seen in cultivated fields and borders. One of the few noncultivated plants evident in the Jalul vicinity and on the tell itself.

Season. Perennial; May-Aug.

This common plant has medicinal uses. A strong emetic and diuretic is derived from the seeds which contain the alkaloids harmin and harmalin. It is unpalatable to animals and therefore survives where other species are selected out by grazing.

Tribulus terrestris L.
Small Caltrops
shiqshiq

A hairy herb found creeping on the ground. The flowers are yellow and in the axils of the leaves. The leaves are compound. *Tribulus* has a distinctive fruit with curved thorns on its surface. It attaches itself to animals and is thus transported.

Habitat. Sandy places, fields, and tracksides.

Season. Annual; April-Aug.

Cultivated Plants Of Tell Hesban Area

CUCURBITACEAE
Cucumis sativus L.	cucumber
Cucurbita sp.	squash
Citrullus vulgaris Schrad.	watermelon

GRAMINEAE
Hordeum vulgare L.	barley
Triticum aestivum L.	wheat
Zea mays L.	corn*

LEGUMINOSAE
Cicer arietinum L.	chickpea
Lens esculenta Moench.	lentil
Phaseolus vulgaris L.	common bean
Vicia faba L.	broad bean

MALVACEAE
Hisbiscus esculentus L.	okra

ROSACEAE
Prunus armeniaca L.	apricot
P. domestica L.	plum
P. persica Stokes	peach

SOLANACEAE
Capsicum sp. L.	pepper *
Nicotiana rustica L.	tobacco *
Solanum lycopersicum	tomato *
Solanum melongena L.	eggplant *

VITIDACEAE
Vitis vinifera L.	grape

* Introduced species

Summary

In summary, the substantial list of local species with significant potential uses must be considered fairly impressive, especially when considered in light of the deteriorated appearing state of the local vegetation at present. However, as the local vegetation resources are limited in comparison to areas of more moist climates, what is available is more likely to be fully utilized. Thus, although beaten down, the local flora still makes a significant contribution to the economy of the project area.

References

Crawford, P., and LaBianca, Ø. S.
   1976   The Flora of Hesban. *Andrews University Seminary Studies* 14: 177-184.

Zohary, D., and Hopf, M.
   1973   *Domestication of Pulses in the Old World*. Science 182: 887-894.

Chapter Six
ECOLOGY OF THE FLORA OF TELL HESBAN AND AREA, JORDAN

Larry Lacelle

Chapter Six

Ecology of the Flora of Tell Hesban and Area, Jordan

Introduction

This chapter examines the plant ecology of the Tell Hesban area in terms of:
1. characterization of the present vegetation communities;
2. effects of natural ecological forces on the present plant communities; and
3. effects of cultivation and grazing practices on the present plant communities.

The project area investigated corresponds approximately to the area surveyed by Ibach (1976; fig. 19) for archaeological sites. The Tell Hesban project area covers an approximate 10 km-radius around Tell Hesban (fig. 6.1). It includes portions of three major physiographic subdivisions:
1. a portion of the Tranjordanian Plateau (the plateau);
2. a portion of the highlands at the edge of the Jordan River Valley (the hills); and
3. a portion of the wadi system that makes up the east side of the Jordan River Valley (the wadi) (fig. 6.1).

This large a project area was evaluated in order to include the vegetation and related ecological zones from the tell to the hinterlands surrounding it. The assumption is that the ecological characteristics of the region surrounding the tell have had, both in the past and present, a significant influence on the livelihood and general way of life of the inhabitants of Tell Hesban and surrounding settlements.

Fieldwork for this paper was carried out in the summer of 1979 in conjunction with fieldwork to collect data to characterize this area's plant species, geology, surficial geology, soils, surface and groundwater resources, cultivation and grazing practices, and population distribution. The success of this field program was greatly aided by the kind assistance of Dr. Andnad Hadidi, Director General of Jordan Department of Antiquities, and his staff members Nadia Razmuzi and Samir Gishan. Fieldwork consisted of numerous walking traverses to familiarize the author with the local ecology, collect data and plant specimens. In addition, considerable time was spent researching technical literature in government and university libraries in Amman.

Specific map units with characteristic geology, surficial material, soil, climate, water and plant relationships were identified and delineated on 1:25,000 scale topographic maps. The units defined are readily recognizable on the ground and reoccur in consistent patterns over the terrain. They have been defined as ecological units as the pertinent ecological factors influencing each of them were considered in its definition. The defined ecological units are described in this chapter and are depicted in a summarized form in figure 6.1. More detailed data on bedrock and surficial geology, soils, surface and groundwater resources, and plant species of Tell Hesban and its surrounding area are contained in Chapters Three, Four, and Five.

General Characteristics of the Flora

Vegetation Zonation and Potential Climax Species

The physiographic subdivisions of the Tell Hesban project area are characterized by Zohary (1962) as being in only one vegetation zone, the Mediterranean Woodland Climax Vegetation Zone (Mediterranean Forest and Maquis Territory) (fig. 6.2). The wadis that descend to the floor of the Jordan River Valley below Tell Hesban are characterized as being transitional to the Mesopotamian Steppe Climax Vegetation Zone (Irano-Turanian Territory) (fig. 6.3). Dominant climax species (species able to reproduce themselves continually on a site) characteristic of

102 ENVIRONMENTAL FOUNDATIONS

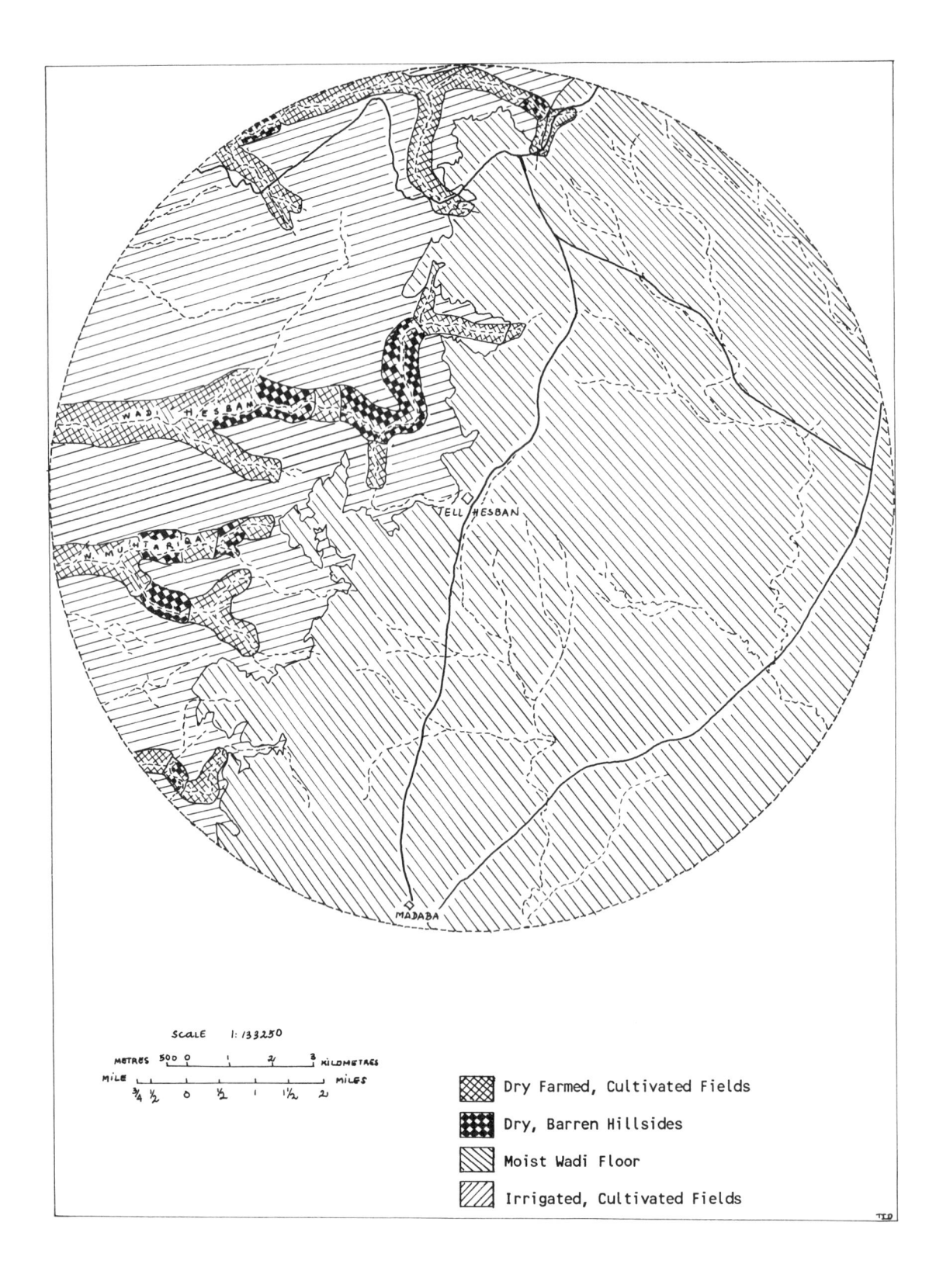

Fig. 6.1 Ecological units of the Tell Hesban project area

ECOLOGY OF THE FLORA 103

Fig. 6.2 Vegetation zonation, Tell Hesban project area

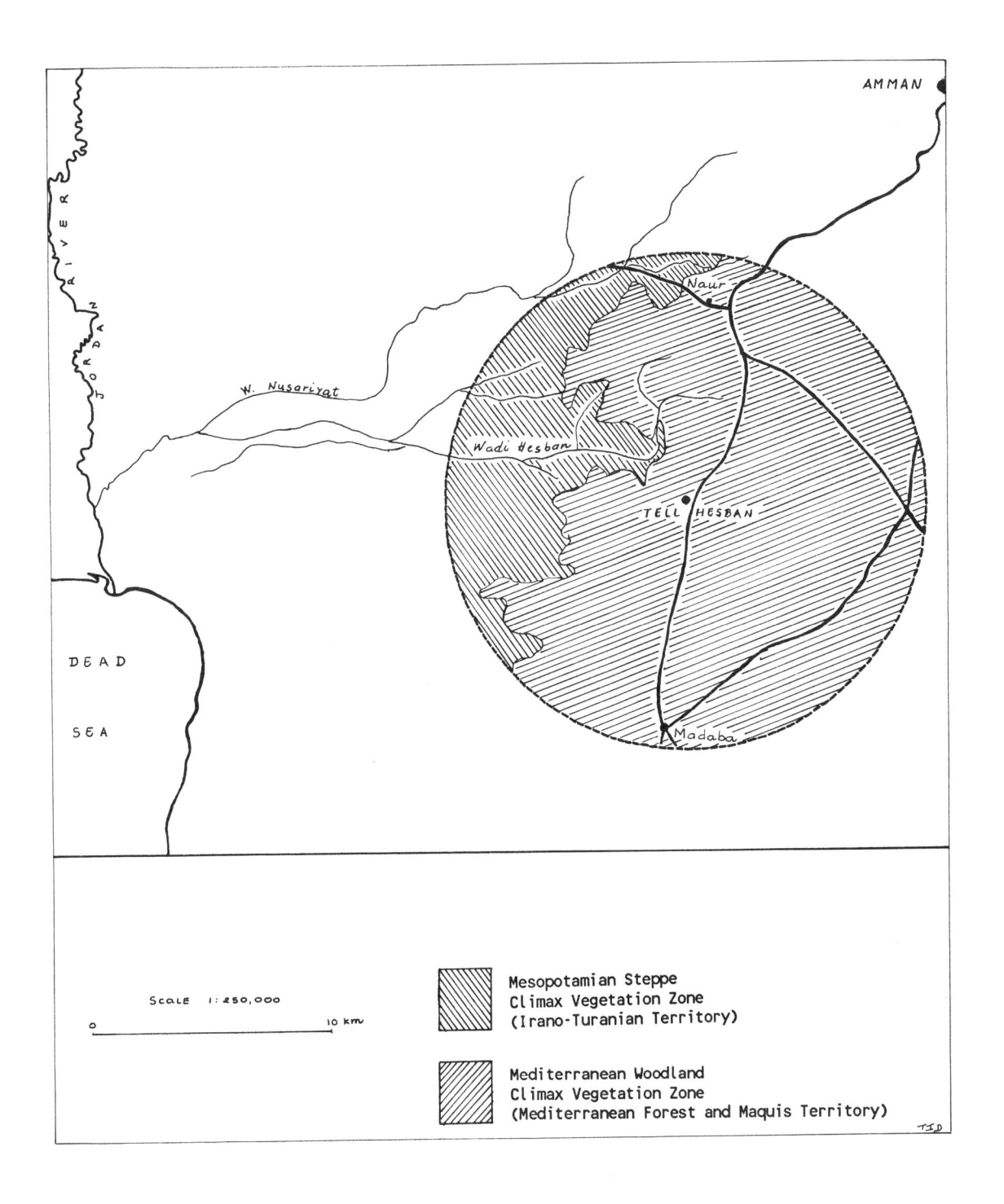

104 ENVIRONMENTAL FOUNDATIONS

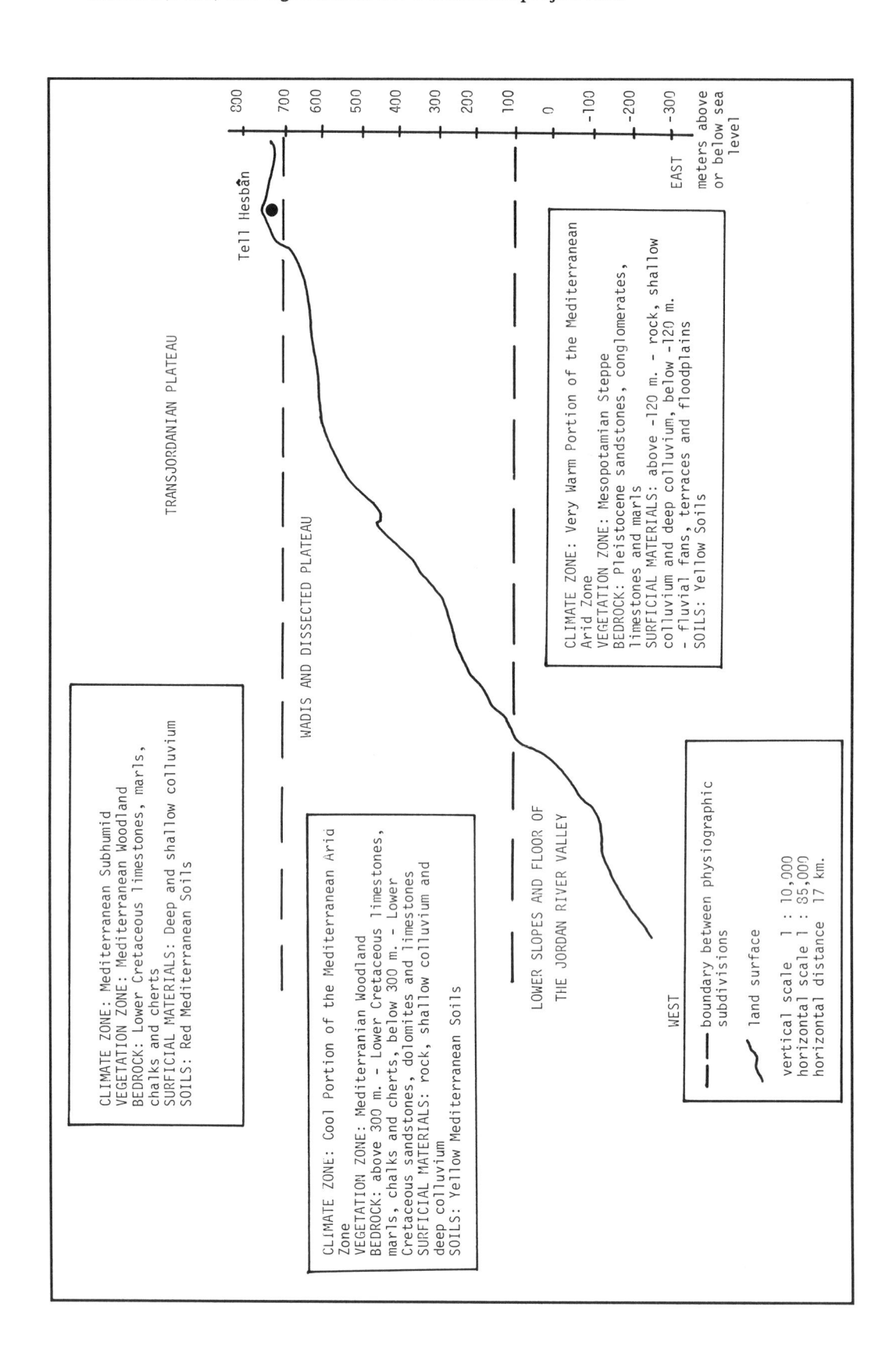

Fig. 6.3   Cross-section depicting elevational relationships of climate zonation, geology, surficial materials, soils, and vegetation in the Tell Hesban project area

the Mediterranean Woodland Climax Vegetation Zone include the oak species *Quercus calliprinos* and the pine species *Pinus halepensis*. A sage, *Artemisia herba-alba,* is a characteristic identifying climax species of the Mesopotamian Steppe Climax Vegetation Zone (Zohary 1962).

Present-day Flora

The present-day nonagricultural plant cover of the Tell Hesban area has little resemblance to the hypothesized climax vegetation (Crawford and LaBianca 1976; Crawford 1986. On the plateau, trees are absent except in orchards or on hilltops that have recently been reforested (pl. 6.1). Aleppo pine (*Pinus halepensis*) and cypress (*Cypressus sempervirens*) are most common in the orchard areas, along with cultivated fruit trees and vines. Species of oak (*Quercus*), pistachio (*Pistacia*) and other species of trees are evident, but not common, in the uncultivated areas. The present-day sparse plant distribution on the lower slopes of the east side of the Jordan River Valley makes this area resemble a desert more than a steppe (pl. 6.2). Plant communities are mainly limited to wadi bottoms, shaded aspects, and localized areas receiving soil water seepage from upslope, or intermittent wadi drainage. The present-day plant cover of the surface of the Transjordanian Plateau, and the wadis of the project area consists primarily of a limited number of spiny and/or unpalatable species that are adapted to withstand both prolonged heat and intensive grazing pressure. The most commonly occurring species of such communities are listed in subsequent sections.

Environmental Parameters and Their
Effects Upon the Flora

Climate

The climate of the Tranjordanian Plateau and the upper parts of the wadis in the Tell Hesban project area is characterized by cool, wet winters (which may include limited periods of frost, snow, sleet, or hail) and hot, dry summers, often resulting in a prolonged summer drought (Chapter Two). The climate on the lower slopes above the floor of the Jordan River Valley is characterized by warmer, frost free winters, and a prolonged, very severe summer drought. The relative mildness of the winters in the Tell Hesban area is suitable for plant species which start growth in the winter and complete their reproductive cycles before the summer drought begins. The virtual absence of precipitation during the summer months eliminates any plant species not able to tolerate prolonged soil water deficiencies, high temperatures, and low relative humidities.

The plateau region of the Tell Hesban area receives approximately 392 mm of precipitation per year (Chapter Two), which decreases to an estimated average of 150 mm in the western-most portion of the project area. The latter quantity is so low that the annual soil water recharge is only partial. Consequently, very little moisture is retained for plants to use later in the dry periods.

Two types of plant species are able to tolerate the prolonged summer drought conditions characteristic of the Tell Hesban area. As mentioned earlier, one type is species able to complete their reproductive cycles in the spring or early summer (winter annuals and winter perennials). The other type is species that are able to tolerate the xeric (dry) soil moisture and climate regimes while maintaining photosynthetic processes (xerophytes and associated plant species such as phlomis [*Phlomis*], sage [*Salvia*], and centaury [*Centaurea*]).

Plants able to complete their reproductive cycles in the spring or early summer include species which die back in summer, leaving only their seeds, bulbs, or tubers for repropagation when moisture again becomes available. Evidence of species of the lily family (*Allium*) was encountered during the field work. Their leaves had presumably been grazed off, but the bulbs left intact. Seed producing annuals are especially well-adapted to the climate regime at Tell Hesban, as their seeds can weather very severe droughts and produce luxuriant, dense growth soon after precipitation has wet the soil. Since sample collection was done during the hottest and driest part of the summer, examples of winter annuals and perennials are undoubtably under represented in the observed species list (Chapter Five).

As mentioned previously, the combination of climate conditions that include high temperatures, low relative humidity and severe soil water deficiencies in the summer, can severely desiccate plant species. Thus, plant species able to limit their transpiration during the drought period are favored in the local plant commu-

Plate 6.1 Reforested hilltops

Plate 6.2 Typical spiny, unpalatable plant species adapted to prolonged grazing pressure

nities. One way that some plants do this is to shed their large winter leaves and replace them with smaller leaves that are still sufficient to maintain photosynthesis yet are able to reduce transpiration significantly. Many plants dominant in the *batha* and *garique* communities (low and dwarf shrubs), reduce transpiration in this way. Thorny burnet (*Poterium*) and headed savory (*Thymus*) are two of the locally dominant species which do so. Another way that a plant under stress can conserve moisture, is to die back so that its limited moisture reserves can be utilized by a few branches at the sacrifice of others. Shrubs which are summer deciduous (lose their leaves in the summer) such as *Lycium* and spurge (*Euphorbia*) are examples of this type of adaption.

Some plant species adapt to the lack of summer precipitation by absorbing moisture from night dew through their leaf surfaces. Subhumid areas in Palestine receive up to 55 mm of dewfall precipitation per year (Trewartha 1968), while desert areas may receive 26 to 37 mm per year (Evanari *et al.* 1971). As much of the dewfall occurs during the drought season, it can be an important component of some plant's water budgets.

Some herbs and shrubs adapt to the summer drought by developing extensive, but shallow, root systems. Thus, a significant proportion of the soil moisture from a rainfall can be absorbed as it soaks into the soil. Such species commonly occur on dry hillsides and are characteristically sparsely distributed, as each plant requires an extensive root area. Plants with a rhizomous root system (systems where the roots spread laterally underground and often propagate new plants) are most successful at this method of moisture collection. Many types of perennial grasses, such as lovegrass (*Demostachya*) and Bermuda grass (*Cynodon*) are of this type. Certain trees and shrubs have extensive shallow lateral root systems also. Pines (*Pinus*) and camel thorn (*Alhagi*) are two examples of such plants.

Another mechanism by which plants can tolerate extreme heat, is to cool themselves by transpiring freely. As this requires an abundance of soil water, such species are confined to the wet sites in the project area. The date palm is an example of such a species.

Tiny plants such as algae and lichens respond to climate induced drought by desiccating to air dryness, becoming dormant, and thus becoming insensitive to further heat and dryness. In the project area, such species are seen on the rocks included in the walls of the ruins as well as on the natural limestone formations exposed by erosion.

Geology, Surficial Materials, and Soils

Limestone and associated chalks, marls, cherts and gypsum are the most common type of bedrocks in the Tell Hesban area (Chapter Three) (fig. 6.3). At lower elevations in the wadis to the west, sandstones, sandy dolomites, and shales outcrop over an extensive area. The limestone bedrock in this area is generally capped by a hard, lime-rich deposit locally called *nari* (see the Bedrock Geology, Surficial Geology, and Soils chapter in this volume). *Nari* characteristically forms a relatively unfissured mantle over the underlying bedrock.

The most significant effect of the predominantly limestone bedrock on the vegetation, is that such rocks weather to calcareous (calcium rich, lime-rich) surficial materials and soils. Even in areas where sandstones outcrop, the soils are often calcareous, due to intermixing with calcareous materials from upslope and to downslope seepage of lime rich, soil water.

The *nari* capping on the limestone bedrock outcrops in the project area, being relatively unfissured, may partly prevent water from accumulating in cracks in the rocks where it would be stored for later plant use. As the soils in many areas where *nari* is evident are thin and relatively course textured, there is very little opportunity for soil water storage.

Bedrocks such as limestones, chalks, marls, and calcareous sandstones contain small amounts of salts from the seawater in which they were deposited. As the rocks weather, small amounts of salts are released and through seepage tend to collect in depressions or on lower slopes. There is some salt accumulation in the soils of the westernmost portions of the project area, near the floor of the Jordan River Valley. Such sites are generally occupied by salt tolerating plant species (halophytes). Halophyte species found in the project area include: saltwort (*Atriplex*), camel thorn (*Alhagi*), canary grass (*Phalaris*), brome grass (*Bromus*), and wild rye grass (*Lolium*).

Erosion of the wadis on the flanks of the Jordan River Valley has produced a rough topography that provides areas of increased shade and areas subject to seepage from upslope. Thus, favorable aspects or slope positions may be characterized by plant species less tolerant of drought and heat than are the species found on less shaded and drier sites.

Much of the plateau surface and limited portions of the hills between wadis are mantled by deep, clay-rich soils with favorable water retaining properties. These soils are also generally characterized by compact soil structure with well-defined cracks between aggregates of soils. These cracks are exploited by plant roots to enable them to reach deep into the soil to extract moisture. These clay rich soils are able to retain soil water until part way through the dry season, and thus enabling plant species requiring the warmer part of the year to complete their reproductive cycles, to survive in this area.

Soil water conditions are much less favorable on soils recently weathered from the *nari*, on soils weathered from sandstones, or on the gullied sandstones and conglomerates characteristic of the westernmost portion of the project area. Such soils are commonly coarse textured, shallow, and droughty. Only plant communities able to tolerate prolonged drought can colonize such sites.

Narrow bands of fluvial gravel, sands, and lesser clays and silts are present on the floors of the wadis. Even during the dry season, the beds of the wadis may carry subsurface flow, or may contain subsurface pockets of water. Thus, plant species able to put down deep tap roots, such as tamarisk (*Tamarix*), camel thorn (*Alhagi*), saltwort (*Atriplex*), and several species of thistles, such as globe thistle (*Echinops*) and cotton thistle (*Onopordum*), can access this underground source of soil water. Because of this extra moisture, the wadi floors are generally characterized by denser and more varied plant communities than are the adjacent slopes (pl. 6.3).

Just north of the Tell Hesban project area, Atkinson *et al.* (1967) describe soils under oak vegetative cover as having an organic matter enriched surface soil horizon and relatively loose soil structure. The soils of the Tell Hesban area likely had such characteristics in the past, but in present times the organic rich surface soils have been destroyed by plowing, natural erosion and, in uncultivated areas, by compaction from overgrazing. Thus, plant species requiring organic matter rich, loose surface soil horizons are at a disadvantage and are likely to be replaced by more hardy species better adapted to degraded sites.

Cultivation

The clearing of the natural plant cover for purposes of cultivation in the Tell Hesban area has definitely been destructive to the local assemblage of plant communities. Plant communities characteristic of the deep, fertile soils of the plateau surface are virtually extinct, as almost every accessible bit of these soils has been cultivated (pl. 6.4). Traverses during the 1979 field investigations revealed that even on rougher topography with shallow, stony soils, nearly every area that could be plowed was being utilized for cereal production. A few local species, able to survive in the exposed environment of plowed fields are commonly found amongst the crops as weeds. Such plants include: plaited leaved croton (*Chrozophora*), spurge (*Euphorbia*), bindweed (*Convolvulus*), camel thorn (*Alhagi*), and blue eryngo (*Eryngium*). Other plant species such as Holy thistle (*Silybum*), centaury (*Centaurea*), cotton thistle (*Onopordum*), hawksbeard (*Crepis*), and sow thistle (*Sonchus*) can be found on rougher topography between fields, but even here climax communities are not represented due to grazing damage and the deforestation damage inflicted by man.

Grazing

Observation of the present-day deteriorated state of the vegetation in the wadis and on the hills of the project area, in addition to observation of thousands of sheep and goats on the range and in the cultivated fields, strongly suggests that the degradation is due to overgrazing (pl. 6.5). The elimination of the expected climax flora of oak and pine species (Zohary 1962, 1973) is no doubt partly attributable to land clearing, but appears to be primarily due to the relentless, unregulated grazing pressure that has not permitted the climax species to regenerate. Succulent, nutritious species have, by being constantly consumed to their roots, been eradicated, leaving the site to thorny and unpalatable species that are able to survive in degraded habitats.

During the field investigations, sheep, goats, and donkeys were commonly observed grazing apparently unpalatable species—mostly thistle-like plants such as: Holy thistle (*Silybum*),

Plate 6.3 Denser and more varied vegetation on wadi floors

Plate 6.4 Utilization of areas of shallow stony soils for cereal crops

centaury (*Centaurea*), blue eryngo (*Eryngium*), and camel thorn (*Alhagi*), as there appeared to be very little other forage available (pl. 6.6). Another deleterious result of overgrazing is that annuals may not be able to become re-established as their seeds have all been consumed. Thus, species that are largely unpalatable, such as common peganum (*Peganum*) and spurge (*Euphorbia*) may come to occupy the site. Other species able to tolerate overgrazing are those that propagate by underground rhizomous stems, such as many species of grasses. Annuals that are able to germinate, grow, and disperse their seed in a short period of time are more likely to survive than those which grow more slowly.

Ecological Units of the Tell Hesban Area

During the traverses of the 1979 field season, it soon became apparent that in the Tell Hesban project area there were definite recurring patterns of ecosystems or ecological units. These ecological units were characterized by similar patterns of: geology, surficial geology, soils, cultivation, grazing, and floral characteristics. The dominant features utilized to characterize the ecological units for this study were similarity of:
1. topographic pattern;
2. depths and textures of surficial materials;
3. the presence/absence of surface and subsurface water; and
4. species and distribution of the flora.

However, the flora of the Tell Hesban area today is so degraded from centuries of man's influence, that areas of more favorable habitat (such as the deep, nutrient rich soils on the plateau) are characterized by plant communities similar to those found in areas consisting of shallow, stony, sandy, drier soils. As a consequence, current plant community patterns were not necessarily the primary factor used to characterize the ecological units, as they do not adequately represent the potential of the environment.

In an effort to determine the relative extent of the defined ecological units, and to subsequently be able to make quantitative inferences as to the effects of such units on the lives of the inhabitants of the area, the areas of the various ecological units were delineated on 1:25,000 scale topographic maps. Unit boundaries were determined both on the basis of field observation and topographic landform as seen on stereoscopic aerial photography and on the maps. The boundaries depicted on the maps are considered very tentative as field observation showed that lesser proportions of secondary ecological units commonly were nested inside the boundaries of large ecological units. These smaller units could not all be satisfactorily depicted at a 1:25,000 scale.

For convenience of reference, the ecologically defined map units have been given descriptive names based on the units most prominent characteristics. Thus, units discussed in subsequent sections include:
1. Dry-Farmed, Cultivated Fields;
2. Dry, Barren Hillsides;
3. Moist Wadi Floors; and
4. Irrigated, Cultivated Fields.

The generalized distribution of these units is depicted in figure 6.1.

Dry-Farmed, Cultivated Fields

These ecological units are primarily characterized by wheat and barley cultivation on the surface of the Tranjordanian Plateau and on hills between wadis west of Tell Hesban (pl. 6.7). These units are extensive in the area studied, making up some 44% of the landscape (fig. 6.1). The relationships between surficial materials, topography, and flora for these ecological units are depicted in figure 6.4. With the exception of irrigated areas and stream banks, the Dry-Farmed Cultivated Fields ecological units are the best suited portions of the project area for plant growth. The soils are typically deep, clay rich, reasonably well-supplied with nutrients and have the capability of retaining soil water for plant use into the dry season. Little of the local natural flora survives on these units, as virtually every hectare is cultivated. In areas of rougher topography between fields, where vegetation disturbance is limited to grazing, present-day seral flora (early stage of vegetation succession on disturbed site) adapted to this ecological unit may be observed. The vast majority of such species are thorny, or unpalatable plants able to tolerate grazing, rather than species characteristic of the assumed climax flora for the area. Tree species of oak, pistachio, pine, and cypress have long since disappeared from the area.

Plate 6.5 View of numerous goats and sheep overgrazing cultivated fields

Plate 6.6 Goat grazing unpalatable thistles

112 ENVIRONMENTAL FOUNDATIONS

Fig. 6.4   Cross-section depicting relationships between topography, surficial materials, and vegetation in the Dry Farmed, Cultivated Fields ecological unit

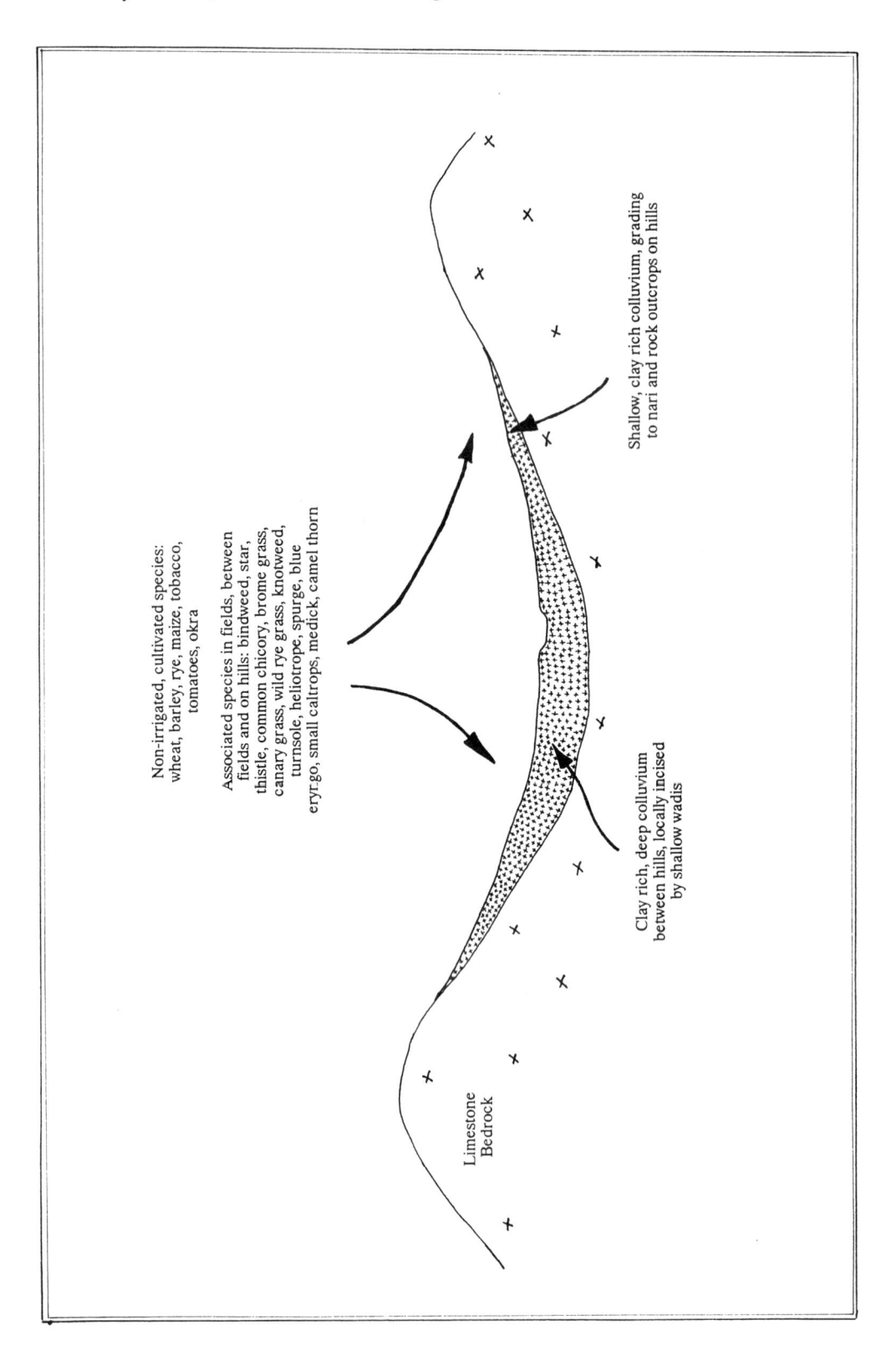

Dry, Barren Hillsides

Areas characterized as Dry, Barren Hillsides generally consist primarily of shallow, stony soils and rock outcrops, but may locally include pockets of deeper soils (fig. 6.4). These ecological units are found on the eroded hills above the wadis, in the rugged, dissected topography of the wadis themselves, and near the summits of hills on the plateau (pl. 6.8). Flora typical of these units may also be seen on the floors of wadis that only flow during heavy rainfalls, but are otherwise always dry. Dry, Barren, Hillside units cover approximately 64% of the project area (fig. 6.1). The relationships between surficial materials, the topography and flora in these units are depicted in figure 6.5.

These Dry, Barren Hillside units are much less suited for plant growth than are the Dry-Farmed, Cultivated Fields, described previously. Since these hillside units are generally steep and have shallow, coarse textured soils, they are much less able to retain soil moisture for plant use in the dry season. Despite these limitations, local pockets of deep soil, as well as many areas of shallow, stony soils, are presently being cultivated for cereal crops. The more rugged uncultivated areas are being heavily grazed. As with all other ecological units in this area, cultivation, grazing and deforestation has eliminated evidence of most climax plant species from the local plant communities typical for these hillsides. As in other ecological units, the species present today are primarily distinguished by their ability to tolerate grazing pressure and the unfavorable growth conditions on the hot, exposed, compacted soil and rock surfaces. Many of the thorny plants such as globe thistle (*Echinops*), thistle (*Cirsium*), and Holy thistle (*Silybum*) are found in this type of environment, as well as shallow rooted species such as white horehound (*Marrubium*), wavy ballota (*Ballota*), and alkanet (*Anchusa*).

Moist Wadi Floors

Since the availability of water is the prime limiting factor in determining the composition of the plant communities in the project area, areas that receive perennial stream flow, that have subsurface flow, or that have pockets of subsurface water during the dry season, have distinctive plant communities in comparison to the dry areas (pl. 6.3). The extent of the Moist Wadi Floor ecological units is exaggerated in figure 6.1, as they actually make up less than 1% of the project area. Despite their limited extent, these units, with their lush flora and available water, have a major importance in the project area, especially with regards to grazing and wildlife habitat. The relationships between the topography, surficial materials, flora, and the watercourses in these units are depicted in figure 6.6.

In the flurial channels, and on the terraces of these channels, plant species characteristic of aerated, high water table soils form lush herb and shrub communities. Water loving species such as common reed (*Phragmites*), cattail (*Typha*), and galingale (*Cyperus*), as well as mint (*Mentha*), oleander (*Nerium*), and tamarisk (*Tamarix*) form thickets that mark the courses of the streams.

Irrigated, Cultivated Fields

The Irrigated, Cultivated Fields ecological units are mapped only below the perennially flowing springs in the Tell Hesban project area (fig. 6.1). Despite only making up less than 1% of the project area, these units have high significance due to their very important roles in supporting year-round cultivation of vegetable and orchard crops, as well as providing water and limited forage for grazing animals (pl. 6.9). The relationships between these ecological units, the adjacent Moist Wadi Floor and Dry, Barren Hillside ecological units, and typical local topography, surficial materials, and flora, are depicted in figure 6.7.

All areas of favorable topography in these irrigated units are cultivated, as are many adjacent areas of rough topography and shallow, stony soils. Consequently, the communities of noncultivated plants are confined to areas between cultivated fields. The between field areas in these ecological units appear to be less heavily grazed than are similar areas in other portions of the project area, probably because farmers are intent upon keeping the animals out of their crops. Even in grazed areas, however, the abundance of soil water permits plants to thrive, and in most cases, to quickly replace consumed, or damaged foliage. Similarly, on such moist sites, plant species are better able to re-establish themselves when portions of the

114 ENVIRONMENTAL FOUNDATIONS

Plate 6.7 General view of plateau surface showing that it is nearly all cultivated with very little area left for natural vegetation

Plate 6.8 Shallow stony soils and dissected topography near the summits of hills typical of Dry Barren Hillsides ecological units

ECOLOGY OF THE FLORA 115

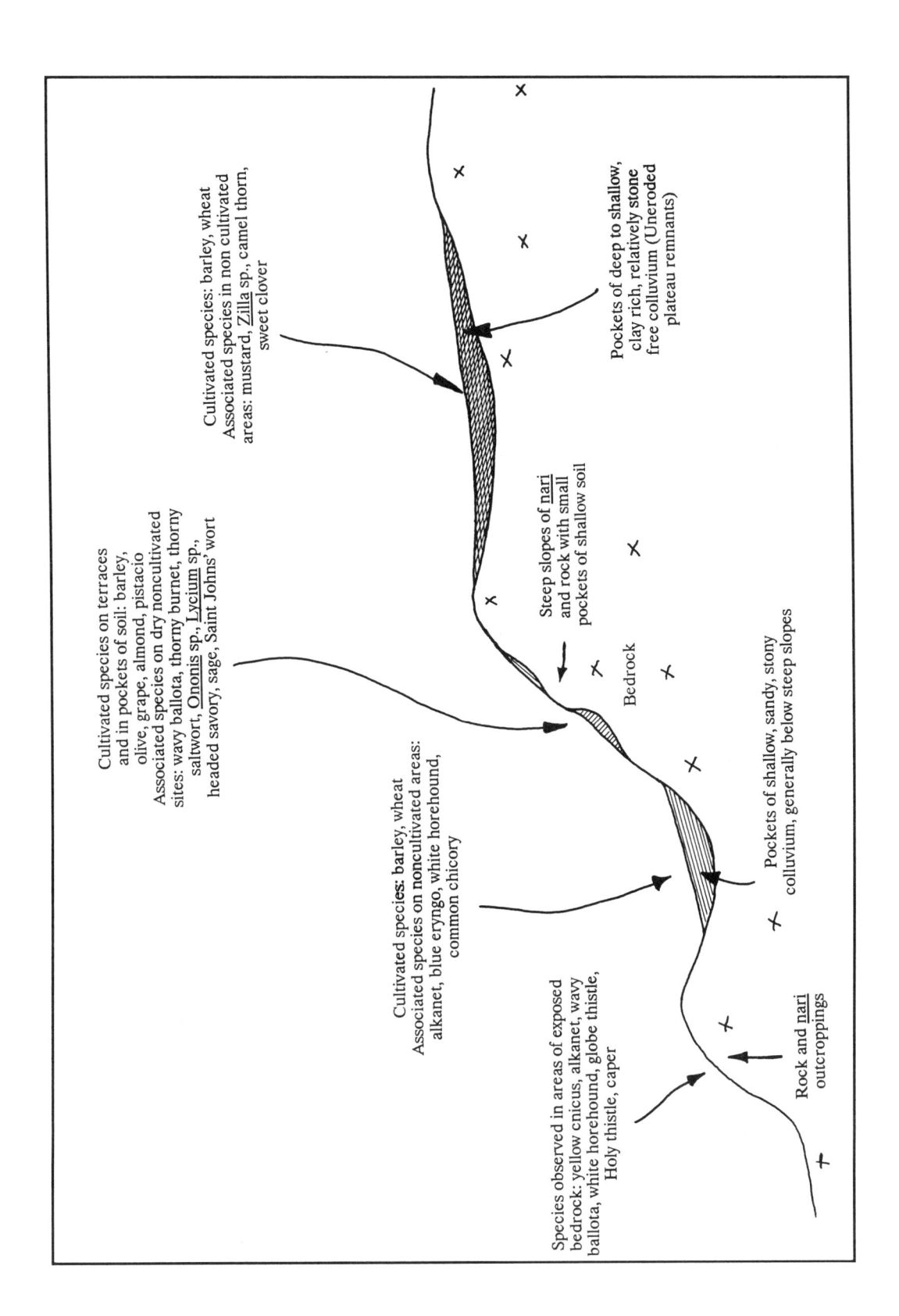

Fig. 6.5 Cross section depicting relationships between topography, surficial materials and plant species in the Dry, Barren Hillsides ecological units

116 ENVIRONMENTAL FOUNDATIONS

Fig. 6.6 Cross section depicting relationships between topography, surficial materials and plants species in the Moist Wadi Floor ecological unit

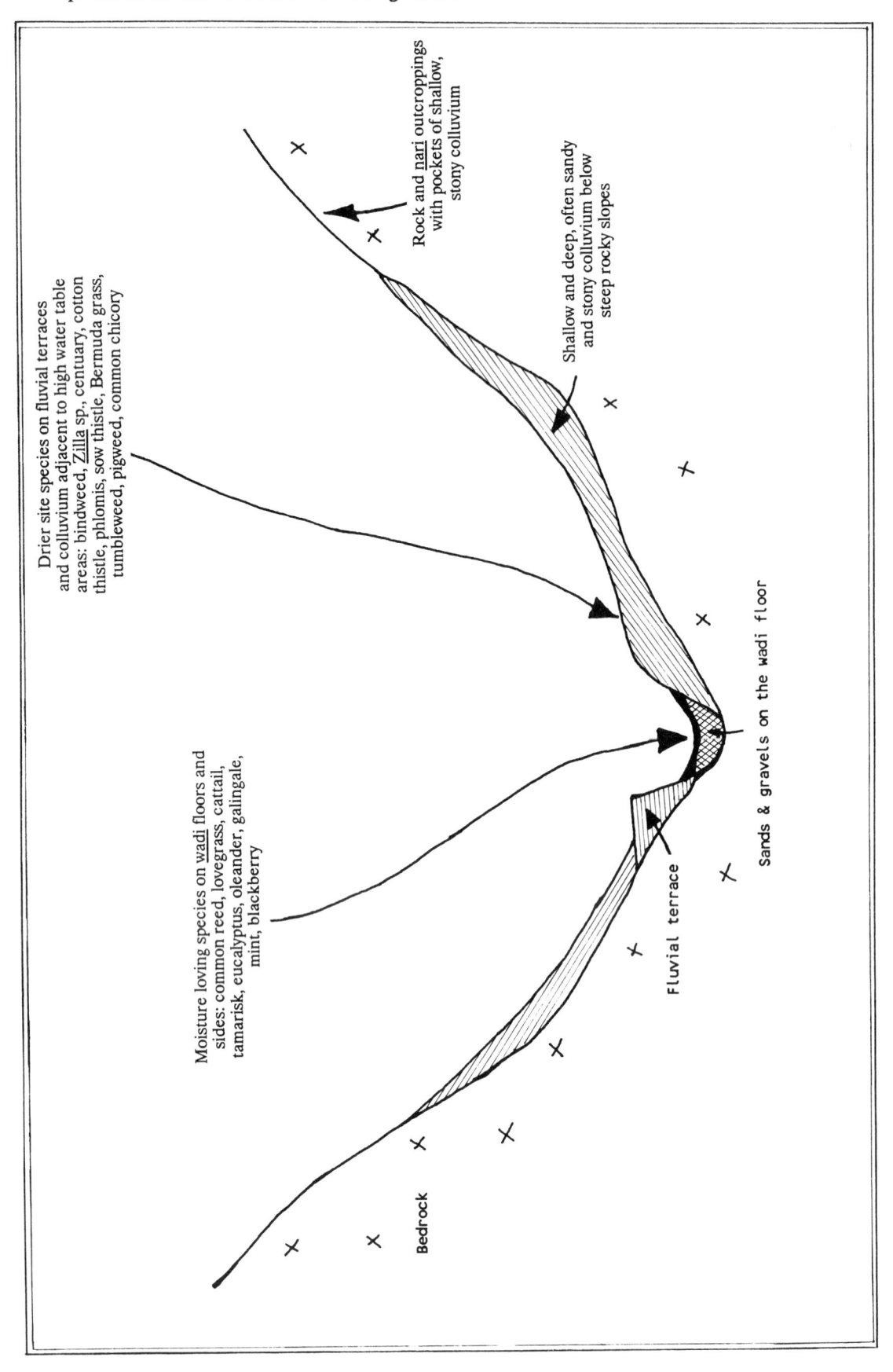

Plate 6.9 General view of lusher vegetation typical of Moist Wadi Floor ecological units

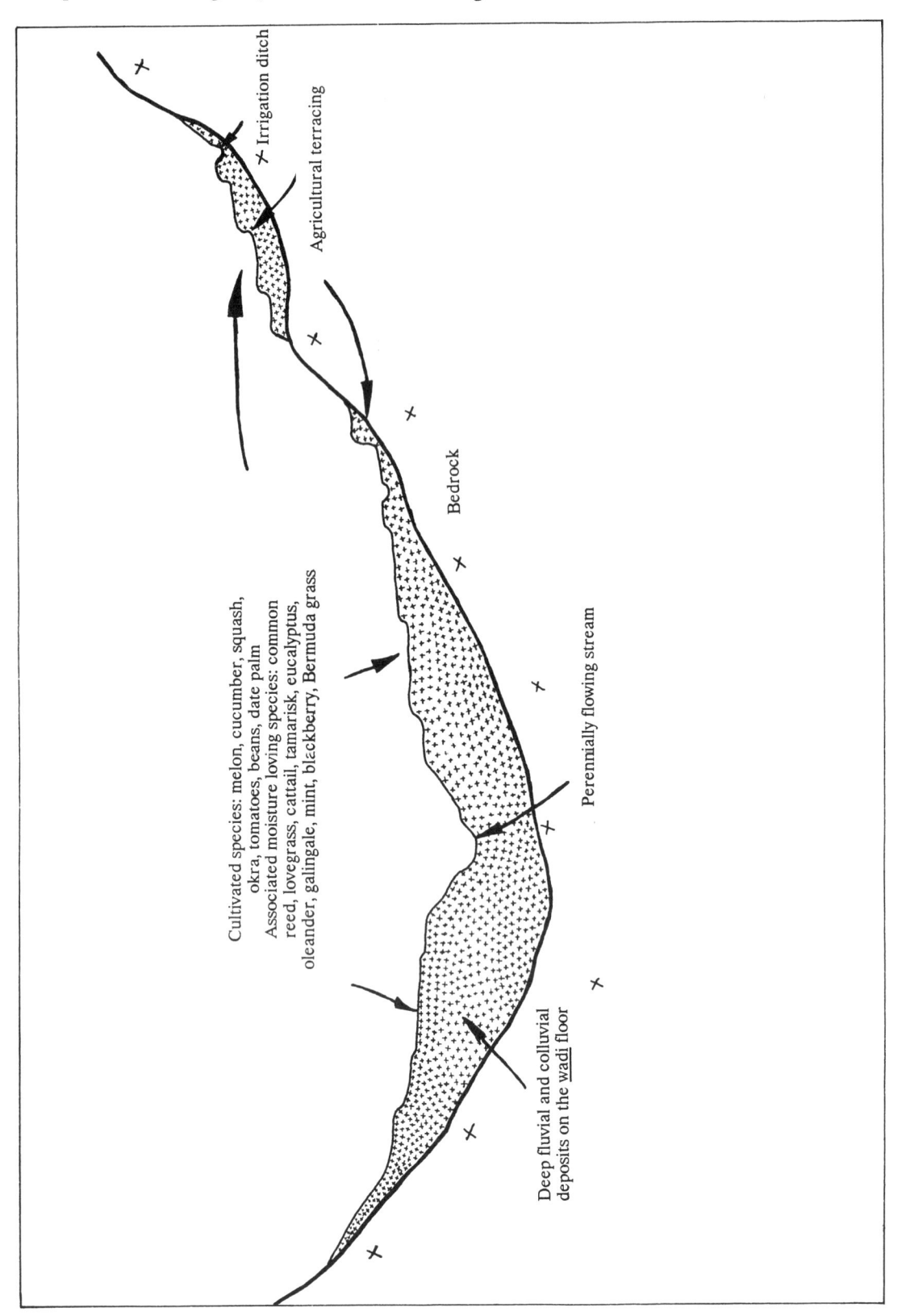

Fig. 6.7   Cross-section depicting relationships between topography, surficial materials, and plant species in the Irrigated, Cultivated Fields ecological units

plant community have been destroyed, in comparison to the previously described dry sites. Nonselective weed species typical for the dry farmed cultivated fields also thrive in these wetter ecological units. Tumbleweed (*Amaranthus*), pigweed (*Amaranthus*), common chicory (*Chichorium*), and *Zilla* were observed in cultivated, irrigated fields. In addition, species described as common in the Moist Wadi Floor ecological units, may also be seen growing between fields in the Irrigated, Cultivated Fields units.

Summary

Observations of vegetation zonation, climate, bedrock and surficial geology, soils, surface and groundwater resources, indicate that the Tell Hesban area has the potential to support a considerably more lush vegetative cover than it possesses at the present time. The fact that species such as oak and pine, which are described as climax species for the area, are for the most part missing from the natural plant communities, indicates that the present-day flora is extremely deteriorated in comparison to its potential species composition, density and distribution patterns. A primary cause of such degradation on noncultivated sites has been centuries of unregulated overgrazing, especially since the project area has become so much more densely settled in recent times. The grazing pressure has so altered the composition of the local flora that vegetation communities on dissimilar habitats, in dissimilar ecological units, often have similar species compositions. As discussed previously, the species that have survived to the present are primarily nonselective, hardy, and generally unpalatable species able to tolerate heavy grazing.

Because of the deteriorated state of the local flora, the definition of specific ecological mapping units in this paper has not been weighted towards identification based on characteristics of the flora. However, despite this, the units defined are sufficiently distinctive in patterns of microclimate, soil, surficial materials, moisture regime, and man's cultivation and grazing activities, to be readily identifiable in the field. It is felt that the units defined in this study represent a valuable basic structure for quantitative discussion of cultivation, grazing, and settlement data for the Tell Hesban project area, both in the present, and in the past. These subject areas are more fully discussed in Chapter Eight of this volume.

References

Atkinson, K. et al.
1967 *Soil Conservation Survey of Wadi Shueib and Wadi Kafrein, Jordan.* University of Durnam.

Crawford, P.
1986 The Flora of Tell Hesban and Area, Jordan. (Chapter Five of this volume).

Crawford, P., and LaBianca, Ø. S.
1976 The Flora of Hesban, A Preliminary Report. *Andrews University Seminary Studies* 14: 177-184.

Evanari, M.; Shanan L.; and Tadmor, N.
1971 *The Negev: The Challenge of a Desert.* Cambridge: Harvard University Press.

Ferguson, K., and Hudson, T.
1986 The Climate of Tell Hesban and Area, Jordan. (Chapter Two of this volume).

Ibach, R. I.
1976 Archaeological Survey of the Hesban Region. *Andrews University Seminary Studies* 14: 119-126.

Trewartha, G. T.
1968 *An Introduction to Climate*, 4th edition. New York: McGraw-Hill, pp. 378-383.

Zohary, M.
1962 *Plant Life of Palestine.* New York: Ronald Press.

1973 *Geobotannical Foundations of the Middle East.* Gustav Fischer Varlag., Stuttgart, Vol. 2, pp. 473-556.

Chapter Seven
PALEOETHNOBOTANY AND PALEOENVIRONMENT

Dennis R. Gilliland

Chapter Seven
Paleoethnobotany and Paleoenvironment

Introduction

In the past, archaeological interest in paleoenvironments has generally centered around the influence of the paleoenvironment on the culture of early man (Castetter 1944). Recently there has been an increasing interest among archaeologists in early man's influence on his environment and his role in the paleoecosystem. Archaeology provides an opportunity to study both ancient human culture and related paleoenvironments simultaneously. In the process of studying how man has lived during a given period of time in the past, the archaeologists may also analyze man's relationship to and effect on his environment during the same time period. Because plants are nonmobile and excellent environmental indicators, botanical remains probably offer the best information on the relationship. With the exception of pollen, carbonized seeds are the most commonly recovered form of botanical remains from archaeological sites.

The purpose of this study was to discover whether any significant change had occurred in the environment of the Tell Hesban area during historical times and, if so, to determine the relationship between such changes and the activities of man. If it is true that desertification in Palestine in general, and around Hesban in particular, has been caused primarily through poor landuse by man, there should be a significant difference between past and present floras, even though the type of physical environment required by both may not necessarily have changed.

Since archaeologists are primarily interested in ancient culture, seeds and other botanical remains have in the past been of importance primarily in indicating whether an ancient society was agricultural and, if so, what crops were grown and for what they were used. The remains of agriculturally important plants have been found at many archaeological sites. Pulses such as peas, lentils, and vetches have been discovered at Jarmo (Helbaek 1960), Jericho (Hopf 1969), Can Hasan (Renfrew 1968, cited in Zohary and Hopf 1973), and Tell Ramad (van Zeist and Bottema 1966, cited in Zohary and Hopf 1973). Remains of fruits such as olives, grapes, dates, and figs have been found at Teleilat Ghassul (Kenyon 1970), Tel Mashosh (Liphschitz and Waisel 1973, cited in Zohary and Speigel-Roy 1975), Lachish (Helbaek 1958, cited in Helbaek 1959), Jericho (Hopf 1969), Tell-ed-Duweir (Schiemann 1953, cited in Zohary and Speigel-Roy 1975), Dikili-Tash (Logothetis 1970, cited in Zohary and Speigel-Roy 1975), and Gezer (Goor 1965, cited in Zohary and Speigel-Roy 1975). Grains such as wheat and barley have been reported from Catal Huyuk (Helbaek 1964), Can Hasan (Renfrew 1968), Hecilar (Helbaek 1970), and Mureybit (van Zeist 1970, cited in Zohary and Hopf 1973).

Based primarily on historical records and literary evidence, Reifenberg (1955) and Mountfort (1964) have stated that Palestine and surrounding areas have been undergoing a desertification process since the time of biblical Israel. Desertification was defined by Rapp (1974, cited in Glantz 1977) as "the spread of desert-like conditions in arid or semiarid areas . . . due to man's influence or to climatic change."

There is a degree of uncertainty about the contribution of possible climatic changes to desertification (Martin 1963; Butzer and Twidale 1966; Lammerts 1971; Cooke and Reeves 1976; Gribbin and Lamb 1978). Based on arboreal pollen percentages from lake sediment cores in northern Israel, Horowitz (1974) concluded that there may have been climatic fluctuations as recently as 1500 years ago, with the rainfall then either being more evenly distributed throughout the year or possible higher by as much as 15-20%. He suggested that the true situation probably involved a combination of the two possibilities. Reifenberg (1955), on the other hand, maintained that there has been no significant change in climatic conditions within historical times.

124 ENVIRONMENTAL FOUNDATIONS

According to Mountfort (1964), Le Houerou (1977), and Kellogg and Schneider (1977), desertification in Palestine and other places has been caused primarily by man. They stated that plowing broke up the root systems which held the soil together and that overgrazing defoliated the rangeland to the point where there was insufficient vegetation to prevent extensive erosion. This process produced the desert-like conditions which are found in much of the Palestine region today.

This question of the importance of man's activities in the origin and spread of deserts has received a great deal of discussion in recent years (Cloudsley-Thompson 1975; Glantz 1977; Secretariat of the United Nations 1977; Went and Babu 1978), but unfortunately, evidence of human influence is somewhat lacking outside of historical records. Such records are usually inadequate as they typically include reference only to the vegetation with an occasional mention of the flora and, more rarely, the temperature or rainfall.

Since 1974, there has been an increasing interest in paleoecology by the Hesban expedition (Bullard 1972; Crawford, LaBianca and Stewart 1976; James 1976; LaBianca 1978). A preliminary list of the modern flora was published by Crawford and LaBianca (1976). Crawford (1986) gives a more extensive list of the modern flora of the Hesban area in Chapter Five of this volume.

By comparing the flora of the past, based on carbonized seeds, with the modern flora, this study examines the effects of man's activities on the environment around Tell Hesban and draws some conclusions about the significance of man's role in the process of desertification.

Materials and Methods

Carbonized seeds were collected at Tell Hesban by Larry Herr during the summer of 1974, by Patricia Crawford in 1976, and Dennis Gilliland in 1978. These were obtained during excavation by water flotation of one 0.25 m³ soil sample from each soil locus, using a 50-gallon barrel of water (Crawford, LaBianca, and Stewart 1976). Since not all of the seeds float, loss of non-floating seeds was minimized by first placing a 1 mm mesh sieve with 15-mm high wooden side walls in the water, and then pouring the soil sample into the floating sieve. This procedure probably did not eliminate the loss entirely, particularly of small seeds. However, since no seeds were found when a sample of the material which had passed through the sieve was examined under a microscope, any loss was probably insignificant. Water is in short supply at Hesban, so each barrel of water was used for flotation of several samples. Contamination between samples was prevented by thoroughly removing the floating debris from the sieve and the surface of the water between samples. After the soil was washed through the sieve, the material remaining on the screen was air-dried and placed in plastic bags to be returned to the laboratory.

A gross sorting was done in the laboratory to remove pebbles and large chunks of charcoal, and the samples were then examined with a stereo-zoom microscope. Seeds were hand-picked from the remaining debris with a small artist's brush and sorted to apparent species. Identifications were based on photographs, drawings, and descriptions of both modern seeds (Fernald 1950; Martin and Barkley 1961; Musil 1963; U.S. Dept. of Agriculture 1971; Renfrew 1973) and carbonized seeds (Helbaek 1959, 1960, 1964, 1970; Renfrew 1973). In addition, some seeds were compared with modern seeds which I collected in the area about Hesban, and some with the private collection of Elwood Mabley of Walla Walla College Place, Washington. Seeds were grouped according to the period from which they were collected, and the percentage abundance of each species present was computed relative to the total number of seeds from that period. Comparisons of species present were made both between periods and with the modern flora.

A total of 80 samples were collected from the Tell Hesban excavations, 19 of which were barren. Of the remaining 61 samples, 6 were from the Iron Age, 8 from the Hellenistic period, 12 from the Roman period, 17 from the Byzantine period, and 18 from the Islamic period. To correct for differences in sample number, the seed data for each period were normalized to a standard sample number of 18 (the number of samples from the Islamic period). This was done by using the equation $N=a(s/18)$ in which "a" is the actual number of seeds for a given species and period, "s" is the total number of samples from that period, and "N" is the normalized number of seeds for that species in the period.

Results

Seeds of approximately 130 taxa were recovered from Tell Hesban samples. Of these, 39 have been identified; 10 to species, 13 to genus, and 16 to the family level. Three more taxa have been tentatively identified to the generic level. Since Amaranthaceae and Chenopodiaceae seeds are difficult to separate and have similar growth requirements (Fernald 1950 and U.S. Dept. of Agr. 1971), these two families have been lumped together for the analysis. For the same reasons seeds of Gramineae were not determined beyond the family level, except for the major agricultural species.

Sixteen families were represented in the findings, including the following: Amaranthaceae, Boraginaceae, Caryophyllaceae, Chenopodiaceae, Compositae, Crucifereae, Cyperaceae, Fumariaceae, Geraniaceae, Gramineae, Leguminosae, Malvaceae, Moraceae, Oleaceae, Polygonaceae, and Vitaceae. There were 6 representatives of Leguminosae, 3 identified species of Gramineae, plus 12 additional unidentified species, 4 representatives of Boraginaceae, 2 of Caryophyllaceae, and 1 each of the remaining 12 families. These taxa are listed in table 7.1 and illustrated in pls. 7.1-7.5.

The relative numbers of seeds of both cultivated and noncultivated taxa from each period is shown in figure 7.1. Taxa which may have been either cultivated or wild include *Heliotropium* sp., *Silene cserei*, *Fumaria sp.*, *Trifolium sp.*, and Amaranthaceae/Chenopodiaceae. The raw data (i.e. number of seeds per taxon per sample) for each period are given in table 7.2.

Two of the 16 families, Geraniaceae and Vitaceae, represented by carbonized seeds are not present in the list of the modern flora of Hesban (See Chapter Five). Eight of the 22 species in the remaining 13 families (Gramineae not included) are also absent from the modern list. Eight taxa are present from the Iron Age, 17 from the Hellenistic period, 8 from the Roman period, 13 from the Byzantine period, and 21 from the Islamic period.

The 91 remaining unidentified taxa occur only in low numbers (1-2 specimens per taxon) and represent only 10% of the total number of seeds collected.

Discussion

Comparison with the Previous Study

Species common to both this study and the preliminary report on botanical remains from Tell Hesban (Crawford, LaBianca, and Stewart 1976) are *Hordeum vulgare*, *Triticum aestivum*, *Lens culinaris*, *Vicia faba*, *Olea europaea*, and *Vitis vinifera*. In addition to these, the earlier study recovered one seed each of *Prunus armenica* and *?Cornus* sp. (=*Zizyphus lotus*, R. Stewart, personal communication, 1979) and three seeds of *Phoenix dactylifera*, none of which were found in this study. Since they were apparently uncommon this is perhaps understandable. Nonetheless, the absence of these seeds from my larger sample size from the same dig site illustrates the caution that must be used when interpreting the seed data.

Ethnobotanical and Paleoecological Interpretations

This section briefly discusses how each of the identified seed species might have fit into the paleoenvironment or cultural setting at Hesban.

Triticum sp. (Wheat, pl. 7.1)

T. aestivum was the only species identified, but as some seeds were quite damaged, other species may have been present. *T. aestivum* was in all probability commonly cultivated at Hesban and probably used in bread or porridge. According to Renfrew (1973), *Triticum* will produce two different types of flour depending on the conditions under which it is grown. Low rainfall and low soil moisture and a hot, dry ripening period produce grain which yields a "strong" flour. This flour makes light, porous loaves when baked. Wheat grown under conditions of high rainfall and a cool ripening period produces a "weak" flour which bakes into harder, more compact loaves. If the past climate at Hesban was similar to the present climate, the former flour type would have been favored.

Hordeum sp. (Barley, pl. 7.1)

H. vulgare (cultivated 6-row barley) was the only species identified from this genus, but *H.*

ENVIRONMENTAL FOUNDATIONS

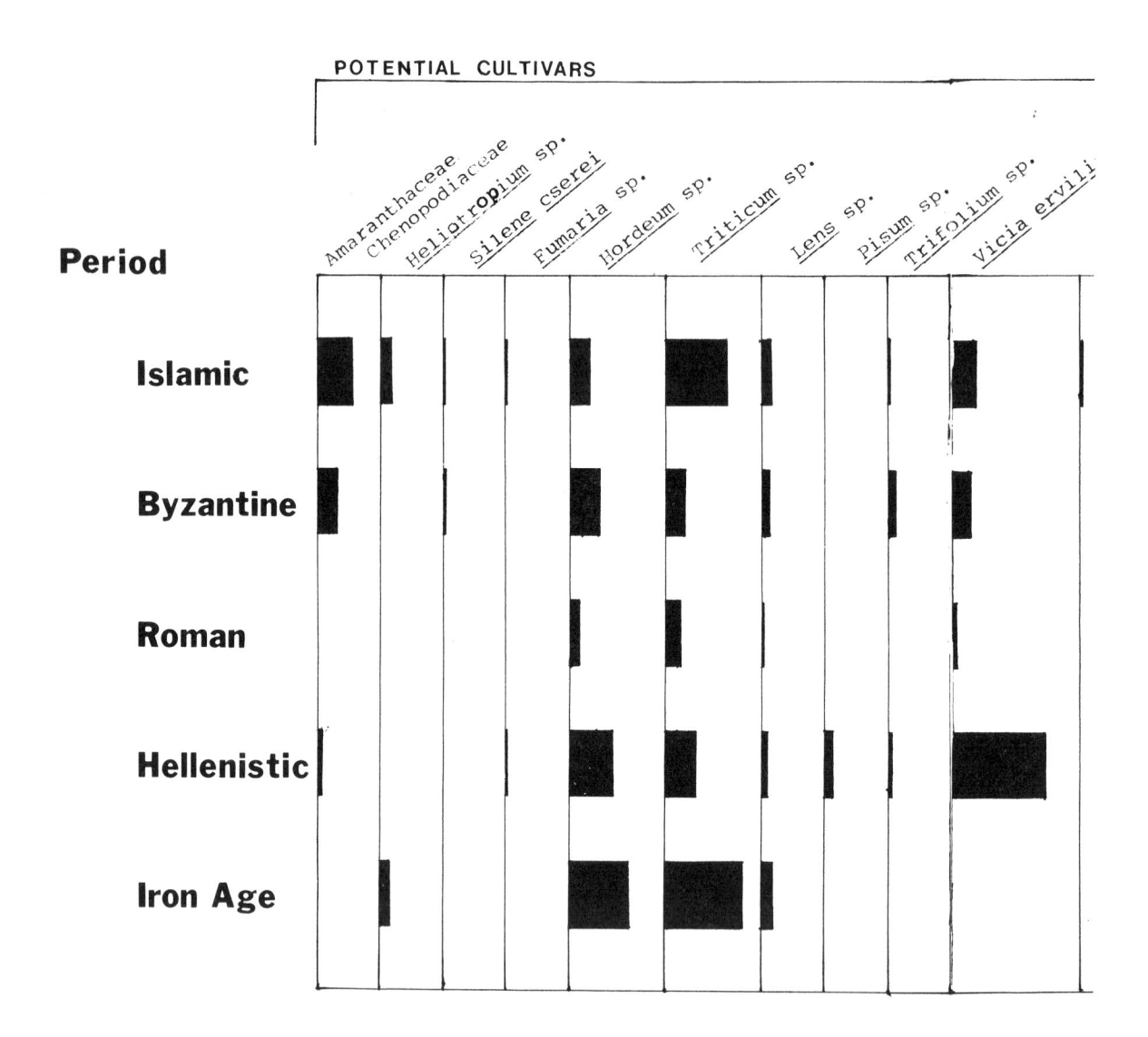

Fig. 7.1   Graph depicting the number of seeds recovered from each archaeological period for each

cultivated or potentially cultivated species and from each of the noncultivated species

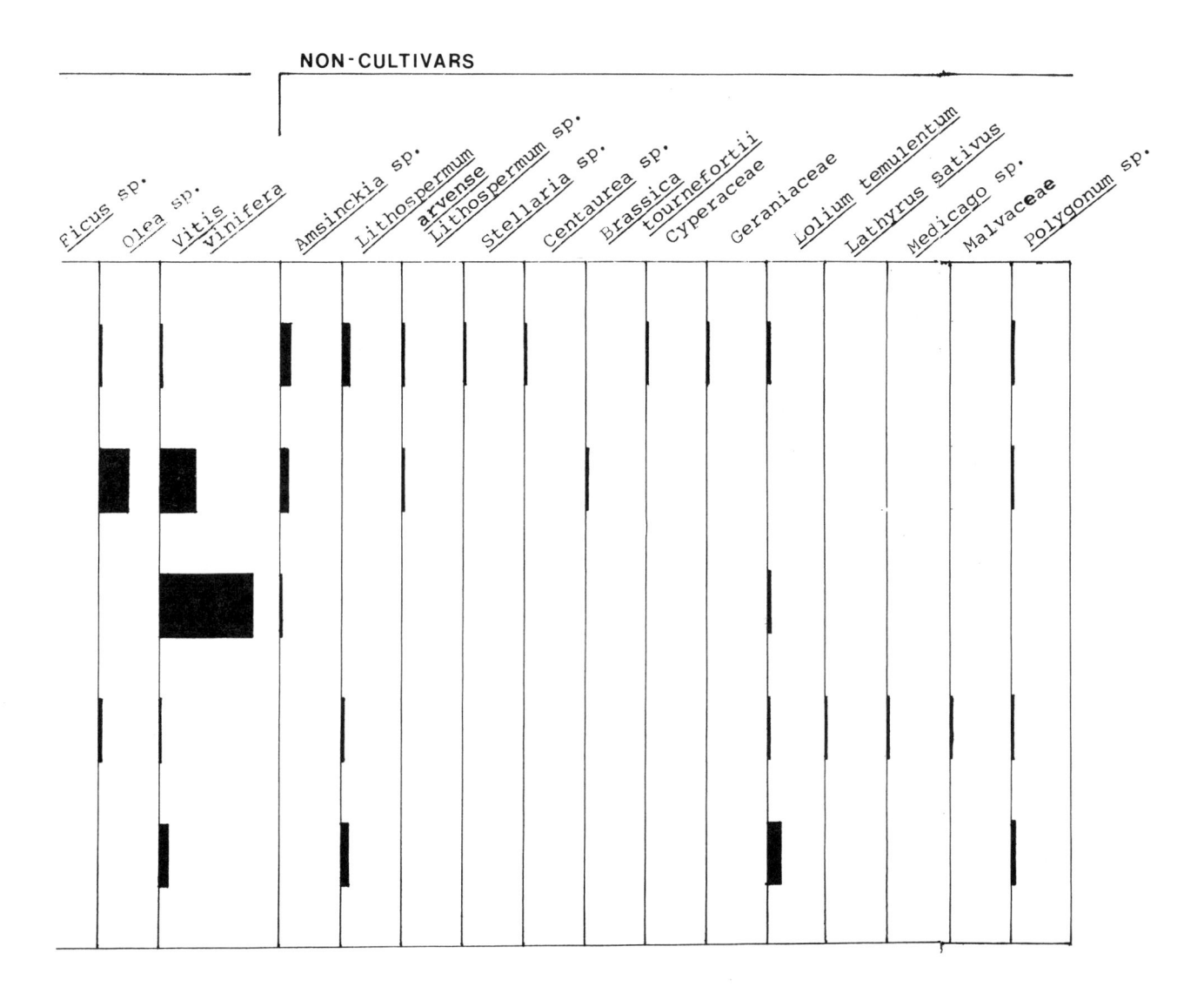

NON-CULTIVARS

Ficus sp.
Olea sp.
Vitis vinifera
Amsinckia sp.
Lithospermum arvense
Lithospermum sp.
Stellaria sp.
Centaurea sp.
Brassica tournefortii
Cyperaceae
Geraniaceae
Lolium temulentum
Lathyrus sativus
Medicago sp.
Malvaceae
Polygonum sp.

Relative Abundance
0 25 50 75 100%

Table 7.1

Table 7.1. A Systematic list of carbonized seeds identified from Tell Hesban, Jordan.

Amaranthaceae	Pigweed family
Boraginaceae	Borage family
Amsinckia sp.	Fiddleneck
Helotropium sp.	Heliotrope
Lithospermum arvense	Gromwell
Lithospermum sp.	Gromwell
Carylphyllaceae	Pink family
Silene cserei	Smooth catchfly
Stellaria sp (?)	Chickweed
Chenopodiaceae	Goosefoot family
Cruciferae	
Brassica tournefortii	Mediterrianean wild turnip
Cyperaceae	Sedge family
Fumariaceae	Fumitory family
Fumeria sp.	Fumitory
Gramineae	Grass family
Avena sp.	Oats
Hordeum vulgare	Six row barley
Hordeum sp.	Barley
Triticum aestivum	Bread wheat
Triticum sp.	Wheat
11 unidentified spp.	Wild grasses
Leguminosae	Pea family
Lens sp.	Lentil
Medicago sp. (?)	Burclover
Pisum sp. (?)	Pea
Trifolium sp.	Clover
Vicia ervilia	Bitter vetch
Vicia faba	Broad bean
Malvaceae	Mallow family
Moraceae	Mulberry family
Ficus sp.	Fig
Oleaceae	Olive family
Olea sp.	Olive
Polgonaceae	Buckwheat family
Polygonum sp.	Knotweed
Vitaceae	Grape family
Vitis sylvestris	Wild grape
Vitis vinifera	Cultivated grape

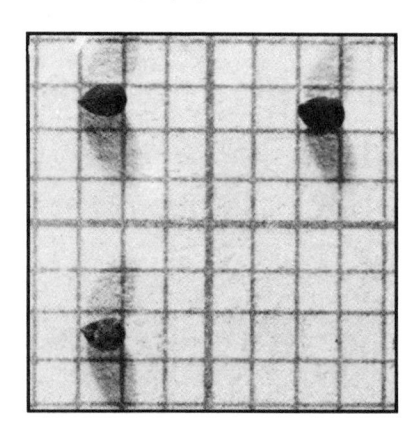

Plate 7.1 Microphotographs of seeds recovered at Tell Hesban

a. *Triticum* aestivum (5x)

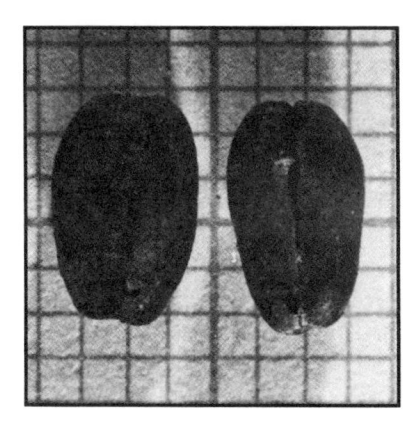

b. *Hordeum* sp. (5x)

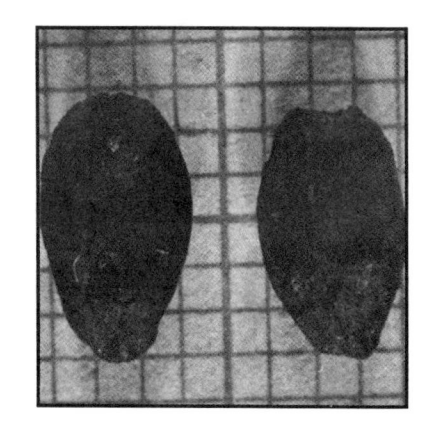

c. *Lens* sp. (5x)

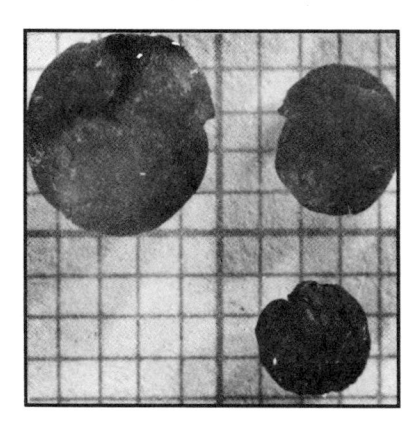

d. *Pisum* sp. (5x)

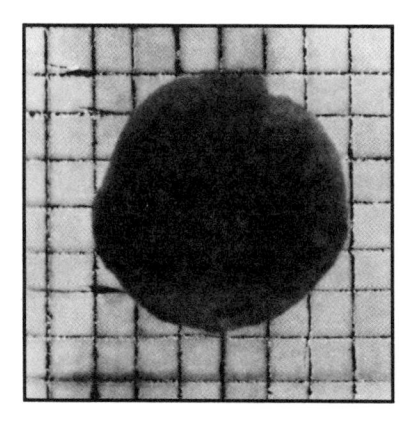

e. *Trifolium* sp. (5x)

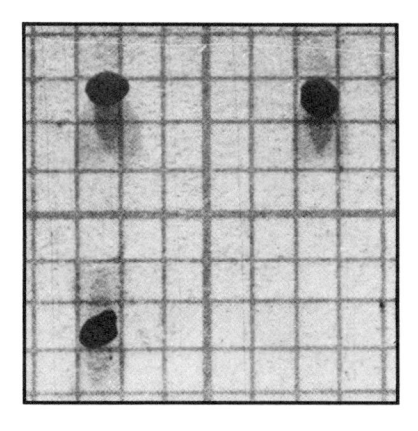

f. *Ficus* sp. (5x)

Plate 7.2 Microphotographs of seeds recovered at Tell Hesban

a. *Vicia ervilia* (5x)

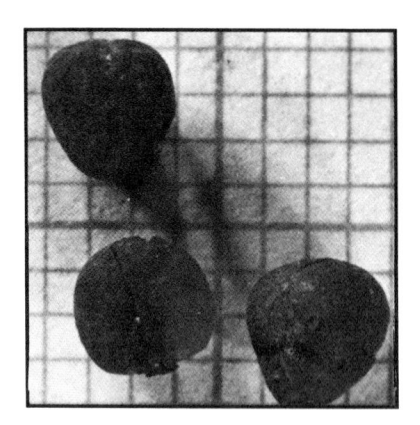

b. *Olea* sp. (2.5x)

c. *Vitis vinifera* (2.5x)

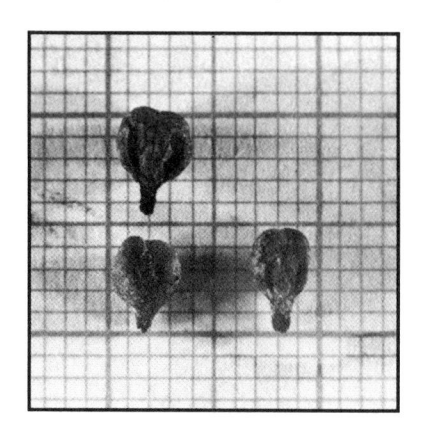

d. *Heliotropium* sp. (5x)

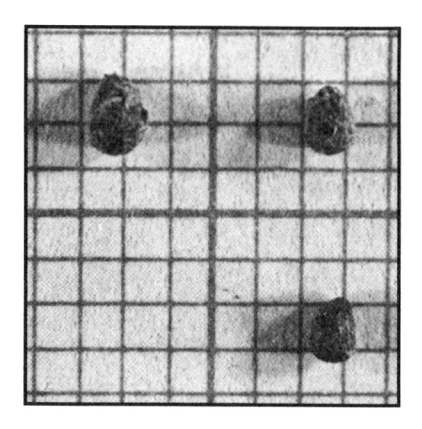

e. *Silene cserei* (5x)

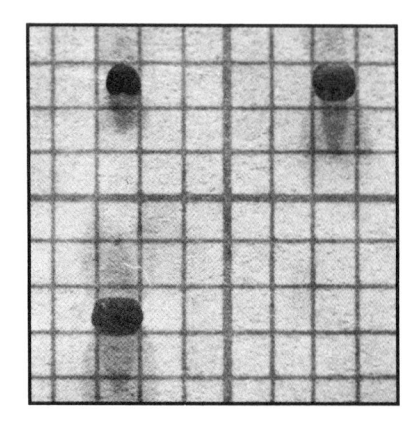

f. *Fumaria* sp. (5x)

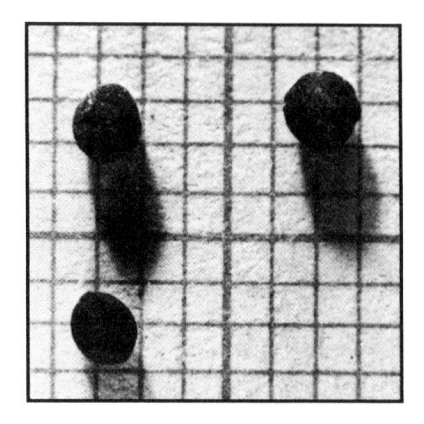

distichon (cultivated 2-row barley) or wild species may have been present among the poorly preserved seeds. *Hordeum* was apparently also cultivated at Hesban and as with *Triticum* probably used in making bread or porridge. It may also have been used to make malt for alcoholic beverages, but there is no evidence to indicate that it was. It must first be sprouted before it can be used in brewing (Renfrew 1973), and none of the seeds collected were in this condition.

Lens sp. (probably *L. culinaris*; Lentil, pl. 7.1)

Lentils may have been cultivated at Hesban, but not as extensively as either *Triticum* or *Hordeum*. It seems likely that it constituted at least a small portion of the diet, however, as it is present in all periods and is cultivated near the modern village of Hesban.

Pisum sp. (Pea, pl. 7.1)

This genus is represented by only one seed from the Byzantine period. Unfortunately, since the specific name is not known, the significance of its presence is difficult to determine. Whether it was wild or cultivated, however, it probably grew fairly early in the spring when rainfall was higher and temperatures were still fairly low (Martin and Leonard 1967).

Trifolium sp. (Clover, pl. 7.1)

More than one species of *Trifolium* was present at Hesban but their specific names have not yet been determined. These species might have been cultivated as fodder for livestock but more likely they were weed species, most of which probably grew along stream banks (wadis) or during moderately cool and wet times of the year such as early spring or late autumn. In either case they would have provided excellent forage and would have been important in nitrogen fixation.

Ficus sp. (figs, pl. 7.1)

The annual rainfall required by *Ficus* is approximately two to three times that suspected at ancient Hesban (Conduit 1947, cited in Renfrew 1973). Thus, this species could theoretically be used to argue for higher rainfall in the past. A doubling or tripling of the rainfall, however, would have increased the probability that less xerophytic species could have out-competed the more xerophytic plants such as some species of *Amaranthus*, and possibly have even eliminated them from the area. More likely, figs were irrigated or imported as dried fruit. Its small leaf area and spreading root system make *Ficus* well-adapted to semiarid conditions (White 1970, cited in Renfrew 1973).

For the semiarid conditions of Hesban cultivation would have been possible under irrigation. The ease with which figs can be dried and stored, however, makes importation an equally good possibility. In addition to being eaten alone, figs may have been used as a sweetener because of their high sugar content (Tibbles 1912, cited in Renfrew 1973). They may have also been used as fodder for domestic animals, especially for pigs (Condit 1947, cited in Renfrew, 1973).

Vicia ervilia (Bitter vetch, pl. 7.2)

One hundred and twenty eight species were collected from the Hellenistic period. Significantly, all but two were found in one sample. According to Helbaek (1970) this species was once a part of man's diet but is now only a fodder crop. This large collection is consistent with the idea that man used it for food. It provides good forage, but is highly susceptible to overgrazing.

Olea sp. (probably *O. europaea*; Olive, pl. 7.2)

Olea is adequately adapted to the climate of modern Hesban and was probably cultivated and utilized there in the past much as it is today. Thus, olives were probably eaten either fresh or pickled. The olive is also valued for its oil content, and so may have been used in cooking and as a lamp fuel (Renfrew 1973).

Vitis vinifera (Grape, pl. 7.2)

Vitis vinifera was probably cultivated at least on a small scale in the past at Hesban. Grapes were likely eaten fresh or dried, or used in making wine (Renfrew 1973). The rather large increase in the number of seeds during the Roman and Byzantine periods (see figure) may indicate that wine was produced on a larger scale during these periods, possibly for export.

Heliotropium sp. (Heliotrope, pl. 7.2)

This plant is today a wasteland species often found in conditions varying from dry, alkaline soils to borders of fresh or saline marshes (Fernald 1950). It is also sometimes cultivated as an ornamental (Lawrence 1951).

132 ENVIRONMENTAL FOUNDATIONS

Plate 7.3 Microphotographs of seeds recovered at Tell Hesban

a. *Amaranthaceae* (5x)

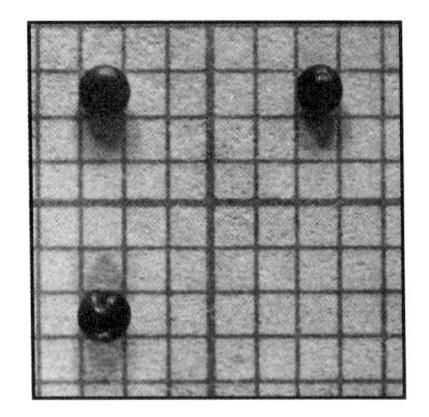

b. Chenopodiaceae (5x)

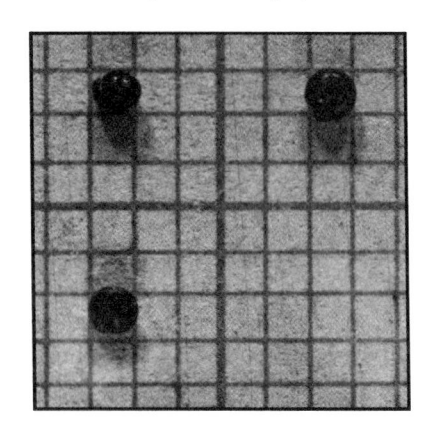

c. *Stellaria* sp. (5x)

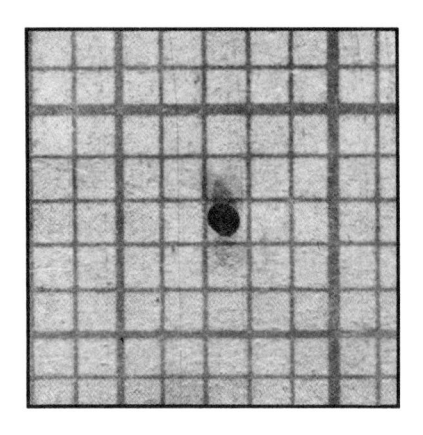

d. *Amsinekia* sp. (5x)

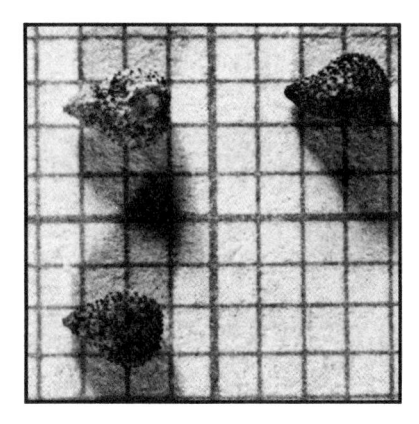

e. *Amsinekia* sp. (5x)

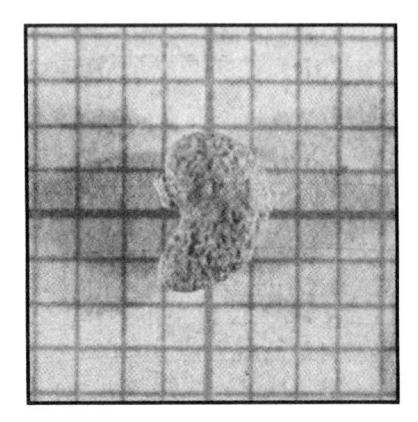

f. *Lithospermum arvense* (5x)

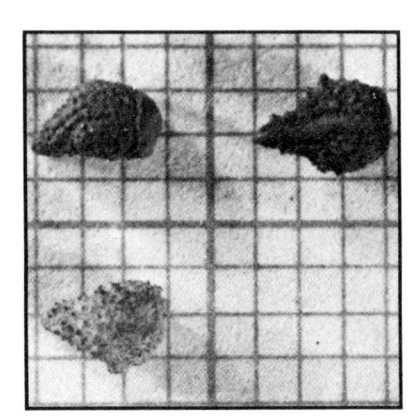

Silene cserei (Smooth catchfly, pl. 7.2)

Members of this genus tend to prefer sandy or rocky soils, though they are often found on cultivated ground. They also seem to favor cool temperatures (Lawrence 1951). This fact could be used to argue for cooler temperatures in the past, but it is not inconsistent with the modern climate. *S. cserei* was likely an early to middle spring species and may have been cultivated as an ornamental, as is the modern plant (Lawrence 1951).

Fumaria sp. (Fumitory, pl. 7.2)

The number of seeds found of this species indicates that it must have been a fairly common weed. It is grown as an ornamental today (Lawrence 1951), however, as it may have been in the past.

Amaranthaceae (probably *Amaranthus graecizans* or *A. retroflexus*; Tumbleweed or Pigweed, pl. 7.3)

Amarnaths are weed species which are frequently found on cultivated soils or wasteland (Lawrence 1951). Since some species do provide edible greens, these plants may have been collected for food or possibly even cultivated (Janick et al. 1974).

Chenopodiaceae (probably *Chenopodium album*; Lamb's quarters, pl. 7.3)

This family is characterized by halophytic and xerophytic species with widespread distribution. They commonly inhabit roadsides, wastelands, and cultivated areas. Some species, such as *Beta bulgaras* (Beets) and *Spinacia olearcea* (Spinach), are cultivated to be eaten as greens, and may have been collected or cultivated in the past. A few species of this family are also ornamental (Lawrence 1951).

Stellaria sp. (Chickweed, pl. 7.3)

The genus *Stellaria* consists primarily of noxious weeds. Its members inhabit a wide variety of habitats from dry to wet, low to high altitude, and woods to meadows (Hitchcock and Cronquist 1973). Thus, no paleoenvironmental information can be derived from its presence at Hesban. A few species provide edible greens (Kirk 1970). Some may also make good forage plants, but they are rarely found on rangeland (Forest Service 1937).

Amsinckia sp. (Fiddleneck, pl. 7.3)

Amsinckia is primarily a wasteland weed (Fernald 1950). It most likely grew along roadsides or footpaths, and possibly around dwellings.

Lithospermum sp. (Gromwell, pls. 7.3 and 7.4)

This genus is represented at Hesban by *L. arvense* and an unidentified species which differs primarily in the shape and size of the hilum. *L. arvense* is a winter annual which grows on cultivated or waste calcareous soils. It is often a troublesome weed in crops of winter wheat (Fernald 1950).

Centaurea sp. (star thistle, pl. 7.4)

This genus is common around modern Hesban but, for obvious reasons, would not have much economic value.

Brassica tournefortii (Mediterranean wild turnip, pl. 7.4)

Brassica contains a number of commonly cultivated species, such as broccoli, cabbage, cauliflower, and turnips. In addition, several species are grown as ornamentals (Lawrence 1951). However, as only one seed was collected, this species was probably a weed, or at best of only minor economic importance.

Cyperaceae (Sedge, pl. 7.4)

Sedges may be found on either wet or dry ground and provide food forage (Forest Service 1937) but the low occurrence indicates that they were probably not cultivated at Hesban.

Geraniaceae (possibly *Erodium* sp.; Filaree, pl. 7.4)

Members of this family are found in a variety of habitats from moist waste places to meadows and woodlands. Some species, such as those of the genus *Erodium*, are valuable range plants (Hitchcock and Cronquist 1973).

Lathyrus sativus (sweet pea; broad bean, pl. 7.5)

This species is represented by only two seeds, both from one Hellenistic sample. This low occurrence indicates that it was only of minor economic importance, if it was used at all by the ancient inhabitants of Hesban. It may however have provided forage for livestock.

Table 7.2 The number of seeds per taxon per sample for each period

Iron Age

Sample #	39	49	52	57	67	78	Taxon Total
Taxon							
Gramineae				13			13
Heliotropium sp		1		1			2
Hordeum vulgare					6		6
Hordeum sp.		1	3	2			6
Lens sp.				2			2
Lithospermum arvense	1						1
Lolium temulentum		1		2	2		5
Polygonum sp.				1			1
Triticum aestivum			2	2	4	4	12
Triticum sp.			1	1			2
Vitis vinifera			1	1			2
Vitis sp.				1			1
Sample Total	1	3	7	26	12	4	53

Hellenistic Period

Sample #	8	15	16	62	71	87	90	Taxon Total
Taxon								
Amaranthaceae/ Chenopodiaceae	2					1		3
Fumaria sp.	4							4
Gramineae	23		1					24
Hordeum vulgare	40	1						41
Hordeum sp.	18		1		1		1	21
Lathyrus sativus	2							2
Lens sp.	8		1					8
Lithospermum arvense	1							2
Lolium temulentum	2							2
Malvaceae	2							2
?*Medicago* sp.	2							2
Olea sp.	6							6
Pisum sp.	1							1
Polygonum sp.	1							1
Trifolium sp.	1		2					3
Triticum aestivum	39							39
Triticum sp.	2				2			4
Vicia ervilia	126		2					128
Vitis vinifera		2						2
Sample Total	280	1	7	2	2	1	2	295

Table 7.2 *Continued*

Roman Period

Sample #	30	40	42	44	48	53	56	66	79	82	86	77	Taxon Total
Taxon													
?*Amsinckia* sp.					1								1
Gramineae	4		1								1		6
Hordeum vulgare	1						1						2
Hordeum sp.				2									2
Lens sp.	1												1
Lolium temulentum										1			1
Triticum aestivum	1		1	1									3
Triticum sp.									2				2
Vicia ervilia	1			1									2
Vitis vinifera	1	1	7	2	3		9		1	9		2	35
Sample Total	9	1	9	6	3	1	10	2	1	10	1	2	55

Byzantine Period

Sample #	4	7	11	12	13	16	6	7	10	14	19	22	25	32	36	50	68	Taxon Total
Taxon																		
Amaranthaceae/ Chenopodiaceae				1	1	1	1			1				5				10
Amsinckia sp.					1										2			3
Brassica tournefortii														1				1
Gramineae			1				4	3		1		1	1	2	1			11
Hordeum vulgare							3	1	1		1	1						7
Hordeum sp.							3	3						1				7
Lens sp.							2	1										3
Lithospermum arvense			1															1
Olea sp.		1		1	2	2	6		1								1	14
Polygonum sp.						1		1										2
Silene cserei	1																	1
Trifolium sp.							2		1									3
Triticum aestivum					2		5								2			9
Vicia ervilia					3		2	3										8
Vitis vinifera					2	1	3				2				7			15
Sample Total	1	1	2	2	11	5	31	12	2	2	2	1	1	9	11	1	1	95

Table 7.2 *Continued*

Islamic Period

Sample #	1	2	5	6	8	9	10	14	17	18	20	2	3	4	5	12	21	23	24	Taxon Total
Taxon																				
Amaranthaceae/ Chenopodiaceae	2	2	3	5	1		20	18	2		2	2					1			57
Amsinckia sp.			1		7			3												11
Centaurea sp.													3							3
Ficus sp.								7												7
Fumaria sp.				2	1			3			1	2								9
Geraniaceae													3							3
Gramineae	4	8	17	3	26		6		1				2	10	6					77
Heliotropium sp.	7									1	1		2		1					14
Hordeum vulgare			2		4		3	6						3						16
Hordeum sp.	1		4		3			2							1					13
Hyperaceae														1	1					1
Lens sp.		1			2		1	11						1				1		15
Lithospermum arvense			3		6	1		1										1		12
Lithospermum sp.			6				1													7
Lolium temulentum		1	1	1			1	2						1	1					3
Olea sp.															1				1	6
Polygonum sp.				1	1		1						1							3
Silene cserei	1																			2
Stellaria sp.	1																			1
Trifolium													1							1
Triticum aestivum		4	8		49		1	21						2						81
Triticum sp.		2	3		15		1	12						5	1					10
Vicia ervilia														3	1	1				36
Vitis vinifera														1	1	1			1	4
Vitis sp.																				1
Sample Total	17	19	49	13	122	1	28	97	2	1	3	3	12	26	14	1	2	1	2	393

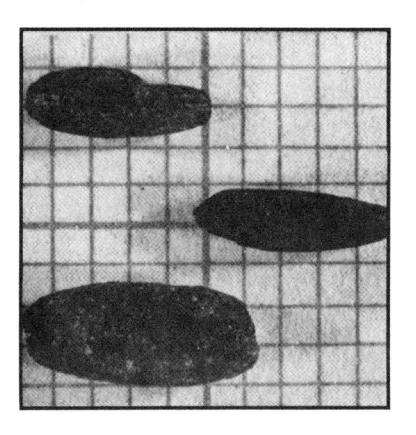

Plate 7.4 Microphotographs of seeds recovered at Tell Hesban

a. *Lithospermum* sp. (5x)

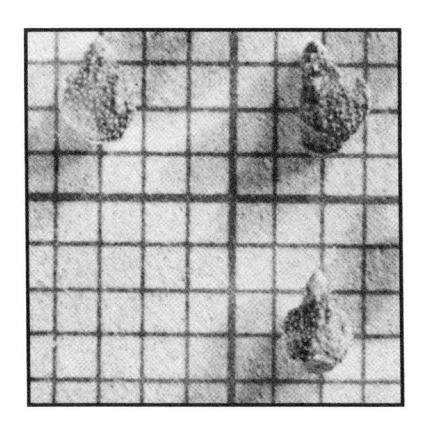

b. *Centaurea* sp. (5x)

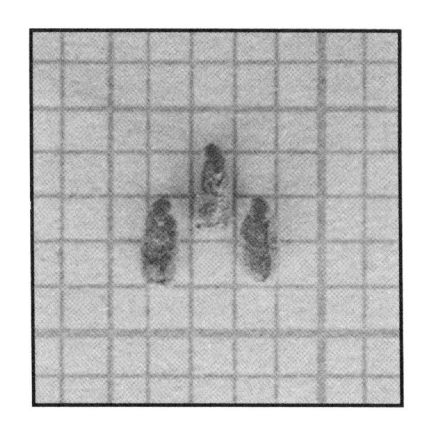

c. *Brassica tournefortii* (5x)

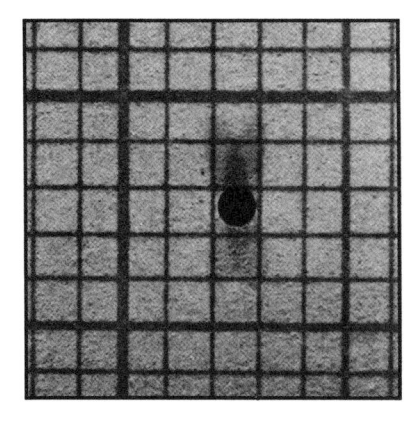

d. Cyperaceae (5x)

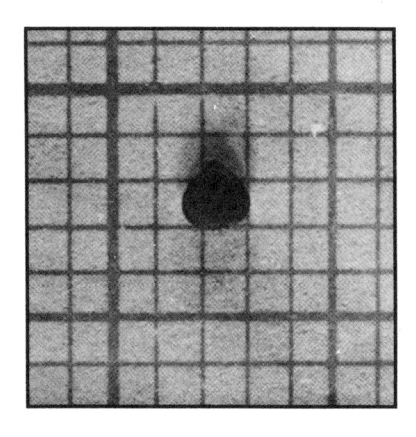

e. Geraniceae (5x)

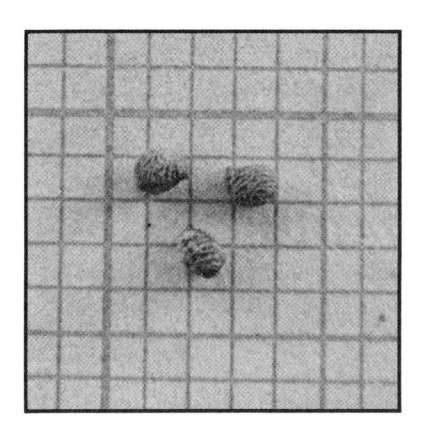

f. *Lolium temulentum* (5x)

Plate 7.5 Microphotographs of seeds recovered at Tell Hesban

a. *Lathyrus sativus* (2.5x)

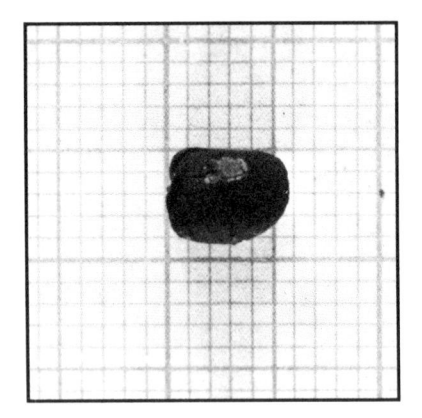

b. *Medicago* (5x)

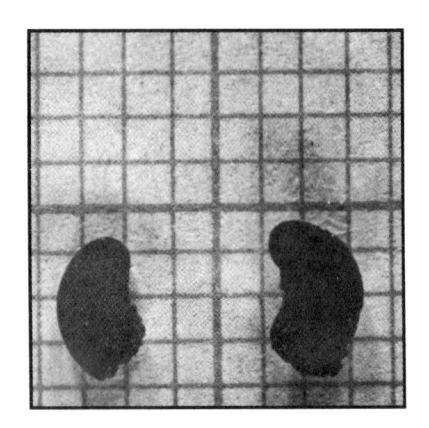

c. Malvaseae (5x)

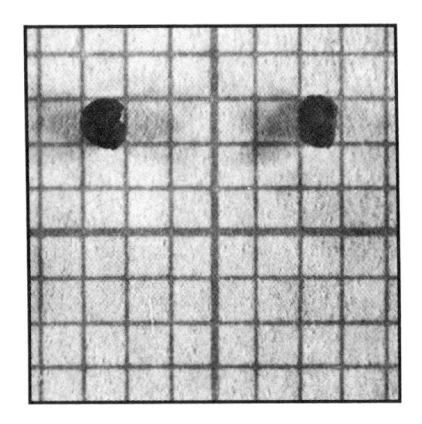

d. *Polygonum* sp. (5x)

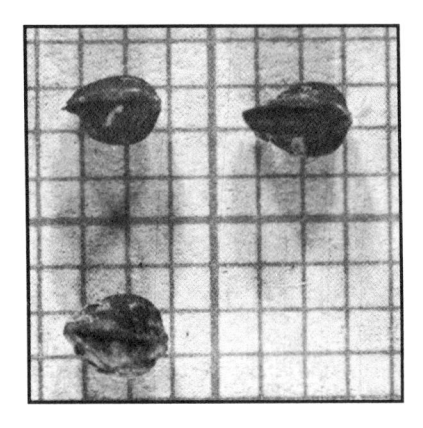

Lolium temulentum (rye grass, pl. 7.4)

This species probably grew as a weed in the Hesban area. It may have provided browse for livestock, but it's low occurrence indicates that it probably had relatively minor domestic importance, if any.

Medicago sp. (Burclover, pl. 7.5)

This species, also, was probably not cultivated since only two seeds were collected. However, it can provide good forage and is important in fixing nitrogen.

Malvaceae (probably *Malva* sp.; Cheeseweed, pl. 7.5)

Members of this genus are low growing weeds often found on wastelands. The presence of these seeds in the Hellenistic period, combined with the fact that member of the same genus are commonly found not far from modern Hesban, suggests that the climate has not changed significantly since the Hellenistic period.

Polygonum sp. (possibly *P. equisetiforme*; Knotweed, pl. 7.5)

The similarity of seeds in this genus makes species identification difficult. These seeds may represent *P. equisetiforme*, one of the modern species of the Tell Hesban area. *Polygonum* is a common weed often found on very poor soil (Hitchcock and Cronquist 1973). It is typical of overgrazed or heavily trampled areas, and has little to no forage value (Forest Service 1937).

Paleoenvironment

The noncultivated plants such as *Stellaria* sp., *Lithospermum arvense*, and *Polygonum* sp., which appear in the botanical remains are more or less opportunistic species. Their presence today in poor soils indicates that they may be colonizing species or part of early successional stages. These observations, combined with their frequent presence on cultivated land, as illustrated by the common association of *Lithospermum arvense* with winter wheat (U.S. Dept. of Agr. 1971), indicate that one thing all the noncultivars have in common is the ability to survive and reproduce under the same conditions as those required by the grain species.

Rainfall requirements for *Triticum* range from 250-1750 mm per year, but more commonly it is grown where the annual rainfall is between 375 and 1150 mm (Martin and Leonard 1967). Germination temperatures range from 3.5-35° with 20-25° C being the optimum range (Peterson 1965). The modern climate of the Tell Hesban area (Chapter Two) corresponds with the lower portions of these optimum temperature and rainfall ranges. Since temperatures are generally lower during the periods of greatest rainfall, conditions are adequate for germination while relatively unfavorable for the development of diseases (Martin and Leonard 1967).

An examination of the habitat requirements of the paleospecies not listed with the modern flora of Tell Hesban (e.g., *Amsinckia* sp., *Fumaria* sp., *Trifolium* sp.) reveals nothing inconsistent with the modern climate. It seems unlikely that a climate change is responsible for their absence today. A year-round study might, in fact, show them to be present as ephemerals or winter species.

Conclusions

What, then can be learned from this study about the relationship between man's activities and desertification at Hesban? Obviously, the limitations imposed by the small sample size, limited area studied, and yet unidentified seeds make it difficult to determine the exact nature of environmental changes, if any. However, two important points are suggested from the data available. First, there has been a change in the flora of the Ḥesbân area during man's occupation. This floral change is documented by comparing the paleospecies with the modern flora. For instance, 9 of the 17 probable weed species present as carbonized seeds are listed in the modern flora. Six of the 8 remaining noncultivated species were present during the Islamic period, indicating that a significant portion of the floral change has occurred fairly recently, as suggested for Palestine in general by Reifenberg (1955) and Mountfort (1964).

The meaning of the floral change, however, is not so clear, and this brings up the second point. The paleobotanical evidence presented here gives no compelling reason to conclude that since the initial Iron Age occupation, the Tell Hesban area has ever been either more or less desert-like than at present. For example, *Amaranthus graecizans* is primarily a desert species (Hitchcock and Cronquist 1973), and either this species or a close relative of it has been present since the Hellenistic period. This situation is also true of *Malva*, as discussed earlier. *Polygonum*, another wasteland weed, was part of the Iron Age flora, and is still present today. On the other hand, legumes such as *Lens culinaris* would not likely grow in a desert without irrigation (Martin and Leonard 1967), and yet this genus has been present since the Iron Age, and even today is cultivated without irrigation. The dominant weedy species of the Hesban area today are primarily thistles, and, since they are adapted to a fairly wide range of habitats, they are more indicative of an early successional sere than of any particular biome. Herein lies our best clue as to what has really happened. It appears that man's activities at Hesban have not caused desertification, but have instead pushed succession backward to an earlier stage. Interpretation of this process as desertification is perhaps an incorrect association of a weed-dominated flora with the unfavorable climate of a desert. The modified flora has come about as a result of the selective advantage given to certain weed species by man's unintentionally selective removal of forage species primarily through overgrazing, as suggested by Mountfort (1964). In some areas on and around the tell, man may have totally removed the native flora through construction and cultivation.

Further study of Tell Hesban seeds may require a significant change in the present interpretations, but on the basis of the evidence presented here, my conclusions are:
1. There is no compelling reason to conclude that the climate of the Tell Hesban area has changed since the initial occupation in the Iron Age.
2. If man's activities at Hesban have caused the observed floral change, that change is not an evidence of desertification, but of a return of the biotic environment to an earlier stage of succession.

References

Bullard, R. G.
    1972   Geological Study of the Heshbon Area. *Andrews University Seminary Studies* 10: 129-141.

Butzer, K. W., and Twidale, C. R.
    1966   Deserts in the Past. Pp. 127-144 in *Arid Lands: A Geographical Appraisal*, ed. E. S. Hills. London: Methuen & Co. Ltd.

Castetter, E. F.
    1944   The Domain of Ethnobiology. *The American Naturalist* 78: 158-170.

Cloudsley-Thompson, J. L.
    1975   Desert Expansion and the Adaptive Problems of the Inhabitants. Pp. 255-268 in *Environmental Physiology of Desert Organisms*, ed. N. F. Hadley. Stroudsburg, PA: Dowien, Hutchinson, & Ross, Inc.

Conduit, I. J.
    1947   Cited in J. Renfrew, 1973. *Paleoethnobotany*. New York: Columbia University Press.

Cooke, R. V., and Reeves, R. W.
    1976   *Arroyos and Environmental Change in the American Southwest*. Oxford: Clarendon Press.

Crawford, P.
    1986   Flora of Tell Hesban and Area, Jordan. (Chapter Five in this volume).

Crawford, P., and LaBianca, Ø. S.
    1976   The Flora of Hesban. *Andrews University Seminary Studies* 14: 177-184.

Crawford, P.; LaBianca, Ø. S.; and Stewart, R.
    1976   The Flotation Remains. *Andrews University Seminary Studies* 14: 185-188.

Fernald, M. L., ed.
    1950   *Gray's Manual of Botany*. New York: American Book Co.

Glantz, M. H.
    1977   The U.N. and Desertification: Dealing with a Global Problem. Pp. 1-15 in *Desertification: Environmental Degradation In and Around Arid Lands*, ed. M. H. Glantz. Boulder, CO: Westview Press.

Goor, A.
    1965   Cited in D. Zohary and P. Speigal-Roy, 1975. Beginnings of Fruit

Growing in the Old World. *Science* 187: 319-327.

Gribbin, J., and Lamb, H. H.
1978 Climatic Change in Historical Times. Pp. 68-82 in *Climatic Change*, ed. J. Gribbin. Cambridge: Cambridge University Press.

Helbaek, H.
1958 Pp. 309 in *Lachish* 4., ed. O. Tufnell. London: Oxford University Press.
1959 Domestication of Food Plants in the Old World. *Science* 130: 365-372.
1960 The Paleoethnobotany of the Near East and Europe. Pp. 99-118 in *Prehistoric Investigations in Iraqi Kurdistan*, eds. R. J. Braidwood and B. Howe. Chicago: University of Chicago Press.
1964 First Impressions of the Catal Huyuk Plant Husbandry. *Anatolian Studies* 14: 121-123.
1970 The Plant Husbandry of Hacilar. Pp. 189-244 in *Excavations at Hacilar*, ed. J. Mellaart. Edinburg: Edinburg University Press.

Hitchcock, C. L., and Conquist, A.
1973 *Flora of the Pacific Northwest*. Seattle, WA: University of Washington Press.

Hopf, M.
1969 Plant Remains and Early Farming in Jericho. Pp. 355-359 in *The Domestication and Exploitation of Plants and Animals*, eds. P. J. Ucko and G. W. Dimbleby. Chicago: Aldine Publishing Co.

Horowitz, A.
1974 Preliminary Palynological Indications as to the Climate of Israel During the Last 6000 Years. *Paleorient* 2: 407-414.

James, H. E.
1976 Geological Study at Tell Hesban, 1974. *Andrews University Seminary Studies* 14: 165-169.

Kellog, W., and Schneider, S.
1977 Pp. 141-163 in *Desertification: Environmental Degradation in and Around Arid Lands*, ed. M. H. Glantz. Boulder, CO: Westview Press.

Kenyon, K.
1970 *Archeology in the Holy Land*. London: Benn.

Kirk, K.
1970 *Wild Edible Plants of the Western United States*. Healdsburg, CA: Naturegraph Publishers, Inc.

LaBianca, Ø. S.
1978 Man, Animals and Habitat at Hesban—An Integrated Overview. *Andrews University Seminary Studies* 16: 229-252.

Lammerts, W. E.
1971 On the Recent Origin of the Pacific Southwest Deserts. *Creation Research Society Quarterly* 8: 50-54.

Lawrence, G.
1951 *Taxonomy of Vascular Plants*. New York: Macmillan Co.

Le Houerou, H. N.
1977 The Nature and Causes of Desertification. Pp. 17-38 in *Desertification: Environmental Degradation in and Around Arid Lands*, ed. M. H. Glantz. Boulder, CO: Westview Press.

Liphschitz, N., and Waisel, J.
1973 Cited in D. Zohary and P. Speigel-Roy, 1975. Beginnings of Fruit Growing in the Old World. *Science* 187: 319-327.

Logothetis
1970 Cited in D. Zohary and R. Speigel-Roy, 1975. Beginnings of Fruit Growing in the Old World. *Science* 187: 319-327.

Martin, A., and Barkley, W.
1961 *Seed Identification Manual*. Berkeley: University of California.

Martin, J. H., and Leonard
1967 *Principles of Field Crop Production*, 2nd ed. New York: Macmillan Co.

Martin, P. S.
1963 *The Last 10,000 Years*. Tucson, AZ: University of Arizona Press.

Mountfort, G.
1964 Disappearing Wildlife and Growing Deserts in Jordan. *Oryx* 7: 229-232.

Musil, A.
1963 *Identification of Crop and Weed Seeds*. Washington, DC: U.S. Department of Agriculture.

Peterson, R. F.
1965 Physiology of the Wheat Plant. Pp. 34-55 in *Wheat: Botany, Cultivation, and Utilization*, ed. R. F. Peterson. New York: Interscience Publishers Inc.

Rapp, A.
1974 Cited in M. H. Glantz, 1977. The U.N. and Desertification: Dealing with a Global Problem. Pp. 1-15 in *Desertification: Environmental Degredation in and Around Arid Lands*, ed. M. H. Glantz. Boulder, CO: Westview Press.

Reifenberg, A.
1955 *The Struggle Between the Desert and the Sown*. Jerusalem: Government Press.

Renfrew, J. M.
1968 Cited in D. Zohary and P. Spiegel-Roy, 1975. Beginnings of Fruit Growing in the Old World. *Science* 187: 319-327.

1973 *Paleoethnobotany*. New York: Columbia University Press.

Schiemann
1953 Cited in D. Zohary and P. Speigel-Roy, 1975. Beginnings of Fruit Growing in the Old World. *Science* 187: 319-327.

Secretariat of the United Nations, ed.
1977 *Desertification: Its Causes and Consequences*. Oxford: Pergamon Press.

Stewart, R.
1979 Personal Communication with author.

Tibbles, W.
1912 Cited in J. M. Renfrew, 1973. *Paleoethnobotany*. New York: Columbia University Press.

U.S. Department of Agriculture
1971 *Common Weeds of the United States*. New York: Dover Publications Inc.

van Zeist, W.
1970 Cited in D. Zohary and M. Hopf, 1973. Domestication of Pulses in the Old World. *Science* 182: 887-894.

van Zeist, W., and Bottema, S.
1966 Cited in D. Zohary and M. Hopf, 1973. Domestication of Pulses in the Old World. *Science* 182: 887-894.

Went, F. W., and Babu, V. R.
1978 Plant Life and Desertification. *Environmental Conservation* 5: 263-272.

White
1970 Cited in J. M. Renfrew, 1973. *Paleoethnobotany*. New York: Columbia University Press.

Chapter Eight
CONCLUSION

Øystein Sakala LaBianca
Larry Lacelle

Chapter Eight
Conclusion

When standing on the summit of Tell Hesban, some 895 m above sea level, a panoramic view is possible of the landscape to the west, to the south, and to the southeast of the tell. Looking westward, one can see the undulating line which forms the edge of the highland plateau, beyond which are numerous hilltops and valleys leading down to the Jordan Valley some 1000 m below. Looking in a southwestward direction one can see the summit of Mount Nebo, and beyond it the northern tip of the Dead Sea, some 20 km in the distance.

To the south and southeast as far as the eye can see, extend the fertile fields of the Madaba plains. Around 10 km to the south, the city of Madaba rises gently above the plain with its conspicuous church steeples and minarets. A few kilometers to the east of Madaba lies Tell Jalul, a prominent mound, clearly visible against the horizon to the east. Except for these two landmarks, the view to the south and east is that of very gently rolling plains, intersected by shallow wadis, winding footpaths leading to isolated hamlets, orchards and scattered villages.

To the north and northeast, the landscape appears more hilly. About 3 km due north the village of El Al, at an elevation of 900 m, interrupts the panoramic circle-vision which otherwise is possible on the summit. Beyond El Al, due north circa 6 km, lies the town of Naur which is situated at the source of the springflow which over the centuries has etched the steep slopes drop down into the bed of the Wadi Naur. East of a line drawn between Hesban and Naur the landscape consists of a plateau in which there are numerous fertile valleys, ranging in elevation between 790 and 860 m, and sometimes steeply ascending hills, some of which rise to over 920 m in height.

While the plains to the south of the tell and the valleys to the north appear as cultivated fields and orchards, the hills and steep walled wadis to the west of the tell appear denuded and barren. In isolated spots, a stand of trees can be seen, the result of recent afforestation efforts by the government of Jordan. For the most part, however, the uncultivated lands not already completely barren appear as low, open scrublands. Surface waters in the form of streamflow cannot be seen anywhere from the summit of the tell except during brief periods immediately following abundant rainfalls.

The research presented in the foregoing chapters provides a basis for a tentative reconstruction of the process of environmental change which has been underway in the vicinity of Hesban since prehistoric times. Fundamental to this reconstruction is the consensus arrived at independently by the various contributors to this volume that macroclimatic conditions have remained largely unchanged since early historical times in this region. In other words, there is no compelling reason to suppose that there have been significant reductions in the amount of rainfall deposited within the project area over the past four millennia. Thus, the climatic conditions which made possible the biotic environment of the past are present also today.

That the biotic environment of the past was significantly more luxuriant and abundant is, however, another point about which there is agreement in the foregoing reports. As Gilliland notes, what accounts for the difference between the somewhat more luxuriant conditions of the past and the degraded conditions of today is a return of the biotic environment to an earlier stage of succession. Herein lie the clues, of course, to our reconstruction of the ancient biotic environment. Our predictions about environmental conditions in earlier times may be made on the basis of established ecological principles regarding the process of change in geological landscapes and biological populations. In our case, of course, we shall draw upon the numerous pertinent insights and examples which have been assembled in the foregoing chapters for this reconstruction.

In times preceding the agricultural revolution's impact on Hesban's landscape, the historical records suggest that a steppe-like vegetative cover, consisting of Mediterranean oak forests interspersed with grasslands, appears to have existed on the plateau surface, on the hills at the edge of the plateau and on the hills of the wadis to the west of Tell Hesban. In the wadis themselves, and in the hills to the north of Tell Hesban, historical records suggest that there were streams and thickets of luxuriant vegetation in watercourses that are today nearly always dry and largely barren. Mediterranean oak forests would have covered most of the hills and slopes in this region, and dense thickets of lush vegetation would have been found along the slopes and in the fertile and well-watered river valleys leading down to the Jordan River. Over the centuries, people's quest for food, shelter, and fuel for fire on this landscape has resulted in gradual removal of the ancient forests. Whether the forests were cut down to clear fields for cultivation, to provide building materials for houses, or to provide fuel for fires; or whether they were prohibited from regenerating as a result of young trees being eaten by grazing herds of animals, their removal had predictable consequences.

To begin with, the gradual removal of the forests and the ploughing of land for cultivation accelerated the natural rates of erosion of soils, especially in the hilly regions. Thus, hills and slopes which once were covered with fertile soils were gradually denuded. As these barren hills and slopes were increasingly exposed to the sun, wind, rains, and man's cultivation and grazing practices, the topography became rougher. As the soils, with their ability to infiltrate and store rain waters, were eroded, surface moisture levels decreased. Rainwater would initiate further gully erosion and then rush at accelerated rates down slopes into tributary and main wadis, and on down into the Jordan River-Dead Sea basin. Increasing penetration of rainwater into the exposed, permeable carbonate bedrocks further decreased the surface moisture conditions.

Soils had their loose, organic matter enriched surface horizons removed by erosion, cultivation or trampling by grazing animals. The soils over time became increasingly compact, and less nutrient rich, factors hindering both cultivation and maintenance of a relatively lush local flora.

Concurrent with these changes in the soil topographic and hydrological conditions were changes in the climate near the ground where surface temperatures on the denuded slopes and plains increased to levels prohibiting establishment and growth of many plant species. Concurrently, with the loss of forest cover, relative humidity levels dropped, eliminating plant species requiring lusher or shaded environments. Plant species able to tolerate hot, droughty conditions and intense grazing pressure, came to be predominant in the local flora. As these species are largely woody, thorny and relatively unpalatable, the forage values of the land were significantly reduced.

Finally, there are good grounds for supposing that the rates at which these changes in Hesban's environment occurred varied significantly over time. Indeed, the process of human alteration of the natural environment must have accompanied the cyclic pattern of intensification and abatement in the local food system over the past four millennia in a predictable manner. Thus, the process of transformation of the local landscape from a luxuriant forest and grassland environment in prehistoric times to what is seen today can be viewed as having occurred in waves, each wave building up during periods of intensification and cresting during periods of food system abatement. We must await the production of other volumes of this final publication series, however, before attempting to delineate in detail the temporal and spatial dimensions of this transformation of Hesban's natural environment.

INDEX

INDEX

Abatement 146
Accelerated climatic processes 21
Achillea 80
Achillea santolina 85
Acreages irrigated 64
Acropolis 84
Adaptation 21, 107
Afforestation 145
Aghanim 90
Agriculture 18, 31, 35, 57
   activities 11, 18, 21
   development 18, 69
   effects of windstorms on 15
   important plant remains 123
   lands 31
   major species 125
   revolution, impact 145
   society 123
   United States Dept. of 45
Agrostis species 89
Ague, cure for 80, 85
Ain
   Hesban 18, 64, 70
   Musa 18, 64, 95
   Sumiya 25, 31, 35, 95
'*Ain el-gott* 90
Air
   circulation 15
   cold 13
   drainage 18
   dryness 107
   flow patterns 13
   lower layers 11
   masses 13
   movements 13, 18
Ait 84
Ajlun 72
   groups 70, 72
Al lig 96
Alcoholic beverages 131
Aleppo pine 80, 94, 105
Algae 107
Alhagi 79, 80, 82, 92, 107, 108, 110
   maurorum 92
Alkaloids 83
Alkanet 113
   prickly anchusa 83
Allium 105
Alluvial plains 88
Almond 79, 95
   trees 93
   oil 96
Altitude 15, 133
Amaranth Family 82
Amaranthaceae 82, 125, 133
Amaranths 82, 133
Amaranthus 79, 119, 131
   graecizans 82, 133, 139
   retroflexus 82, 133
American Mastic 82
Amman 35, 72, 83
Amsinckia 133
   species 133, 139
Anacardiaceae 82
Anchusa 113
   strigosa 83
Ancient
   culture 123
   forests 146
   man 35
   roads 35, 64
   terracing 64
   times 88, 94, 96
   water management 64
Ancients 88
Andrews University 9
Animal shelters 35
Animals 84, 87, 88, 89, 92, 97, 113, 146
   bedouin 95
   grazing 87, 88, 92, 95, 96, 113
Annual soil water recharge 105
Anthropogenic sediment 28
Apocynaceae 82
Appetite stimulus 84
April, 13, 83, 85, 86, 88, 89, 90, 91, 92, 93, 95, 96, 97
'*Aqool* 92
Aquifers 18, 64, 69, 70, 72
   Ajlun 72
   contamination 72
   high transmisivity zones 70
   limestone 61

Aquifers (*Continued*)
　lower 72
　pollutants 72
　salinity 70
　sandstone 64
　stock watering 70
Arabic 28
Arable area, decrease 21
Arabs 86, 91
Arboreal pollen percentages 123
Arid
　areas 123
　climates 15
Artemisia 79, 80, 85
　herba-alba 85, 105
　herba-alba association 85
Artichoke 86
Artist's brush 124
Asia 13
Asian High 13
Asparagus 79
Asparagus acutifolius 92
Asses Fig 93
Asuf 83
Atlantic 11
　depressions 13
　Ocean 11
Atmospheric
　condition 11
　pressures 13
　water 18
Atriplex 79, 82, 84, 107, 108
　species 84
August 16, 18, 82, 86, 88, 90, 91, 92, 93, 95, 96, 97
Australia 84, 93
Australian oak 80, 84
Autumn 84, 131
Avena 80, 89
　barbata 89
'*Awsaaq* 96
Azores 11
　High 15

Bahma 90
Bake ovens 95
Balkans 13
Ballota 91, 113
　undulata 91
Baluchistan 11
Banduret deeb 96
Banks 87, 95

Bark 82, 88, 93, 94
Barley 123, 125
　cultivated 90, 110
　cultivated 2-row 131
　cultivated 6-row 125
　fields 77
　wild 90
Barometric high 13
Barwek 96
Basalt 35
Baseflow 64, 69
Batha 92, 107
Batha-garigue 83
Bedload 64
Bedouin 95
Bedrock 25, 35, 38, 45, 48, 53, 69, 70, 101, 107, 119
　Aljun 72
　aquifers 69, 70
　calcareous 70
　calcareous sandstones 107
　carbonate 31, 35
　chalks 107
　characteristics 28
　cherts 107
　eroded 48
　exposure 35, 70
　fossils 31
　fractures 70
　fragments 45
　geology 25, 31
　geology, influence of man 35
　geology, influence on man 35
　gypsum 107
　history of deposition 25
　Kurnub 72
　limestone 31, 107
　lithified 35
　marls 107
　permeabilities 70
　pockets 38
　porous surfaces 70
　rates of water transmisivity 70
　sandstone 45
　solution pockets 70
　weathering 31
Bees 91
Beets 133
Behavior 21
Beidha 84
Belqa group 70
Bent grass 89
Bermuda grass 80, 82, 89, 107

Berries 82, 89, 95, 96
Beta bulgaras 133
Bible 88, 90, 94, 95
Bilharzia, treatment of 92
Billan 95
Bindweed 87, 108
    Family 87
Biological populations 145
Biome 140
Biotic
    environment 140
        perspective 21
        value 18
Bishreen 85
Bitter gourd 80, 88
Bitter vetch 131
Black nightshade 79, 96
Black scale 82
Blackberry 96
Bleeding, inducement 85
Blood thinner 85
Blue eryngo 79, 97, 108, 110
Bohma 89
Borage Family 83
Boraginaceae 83, 125
Bordi 96
Borealo-Tropical group 82, 84
Botanical remains 123, 125, 139
Boulos 92
Boundaries 110
Branches 82, 83, 84, 85, 86, 91, 92, 93, 95, 97, 107
Brassica 79, 80, 133
    species 87
    tournefortii 133
Bread 125, 131
Brewing, alcohol 131
Britain 85, 86, 91
Broad bean 133
Broccoli 133
Brome 89
    grass 80, 89, 107
Bromus 80, 107
    species 89
Browse, livestock 139
BSk
    classification 15
    climate 15, 16
Buckthorn Family 95
Building materials, for houses 146
Bulbs 21, 79, 105
Burclover 139
Burial 35

chambers 31
Burnet 91, 92
Bus 90
Buttercup Family 95
Byzantine
    church 31
    period 131
    period, seeds 124
    period, taxa 125

Cabbage 133
Cactaceae 83
Cactus 83
    Family 83
Calcareous 45
    duricrust 28
    materials 107
    rocks 28
    sandstones 107
    soils 94, 107
    surficial materials 107
Calcic
    recrystallized 28
    Rhodoxeralfs 48
    Yermosols 48
Calcium 53
    carbonate, precipitation 48
    rich 107
Calcrete duricrust 28
Caliche duricrust 28
Caltrops Family 97
Camel thorn 79, 80, 82, 92, 107, 108, 110
Camels 92
Campbell Stokes recorder 16
Can Hasan 123
Canary grass 82, 90, 107
Canyons 38
Caper 79, 80, 81, 83
Caper Family 83
Capparideaceae 83
Capparis 79, 80, 82
    spinosa 83
Capping rocks 28
Carbohydrates 84
Carbonate
    building materials 31
    enriched 28, 48
    rocks 31, 70
    seeds 5, 123, 124, 125, 139
    strata, resistant 28
    Upper Cretaceous 31
Carbonate bedrocks 28, 38, 61

Carbonate bedrocks (*Continued*)
   penetration of rainwater 146
   upper strata 28
Cardiac glucoside, active 80, 83
Carrot Family 97
Caryophyllace 84
Caryophyllaceae 125
Cashew Family 82
Casuarina 80
   stricta 84, 88
Casuarinaceae 84
Catal Huyuk 123
Catchments 64
   basins 38
Cattail 79, 96, 97, 113
Cattle 82
Cauliflower 133
Caves 28, 35
   sediment 35
Centaurea 79, 80, 105, 108, 110
   hyalolepis 85
   iberica 85
   species 85, 133
Centaury 79, 85, 105, 108, 110
Central Asian High 13
Chalks 28, 70, 85, 107
   crop fields 79
   crops 113
   mining operation 35
   production 108
   substitute 84
Channels 38, 61
   flurial 113
   terraces 113
Charcoal 35, 124
Cheeseweed 139
Chemical
   soil analyses 53
   soil property 53
Chenopodiaceae 84, 125, 133
Chenopodium 79, 82, 84
   album 84, 133
Cherry 93
Cherts 28, 31, 45, 107
   blocks 31
   insoluble 38
   resistant 28
   strata 70
Chichorium 82, 119
   intybus 85
Chichorum 79
Chicken egg 83
Chickweed 133

Chicory 79, 85
China 96
Christian region 83
Chromic Luvisols 48
Chrozophora 80, 108
   plicata 88
   tinctoria 89
Church steeples 145
Cirsium 113
Cisterns 31, 61, 64, 69
Citrullus 80
   colocynthis 88
   species 88
Clay 31, 45, 48, 53, 108
   illite 38, 45
   kaolinite 38, 45
   montmorillonite rich 38
   soils 108, 110
Cliffs 83
Climate 9, 21, 105, 119, 139, 140
   bounds, behavioral response 21
   changes of 5, 9, 123, 139
   characteristic 101
   classification 15, 16
   conditions 9, 13, 105, 123, 145
   data 9
   dry 84, 94
   fluctuations 123
   human activities 21
   impact on humans 21
   impact on vegetation 21
   induced drought 107
   local 15
   modern 131, 133, 139
   moist 98
   normative patterns of change 21
   past 125
   present 125
   regime 9, 13, 15, 105
   soils 21
   unfavorable 140
   warm 88
Climatic
   parameters 16, 21
   variables 15
   phenomenon 15
Climatic indices
   clouds 15
   condensation 15
   evaporation 15
   precipitation 15
   pressure temperature 15

relative humidity 15
standard time 15
sunshine duration 15
vapor pressure 15
wind 15
Climax communities 108
   flora, assumed 110
   flora, elimination 108
   plant species 113, 119
   species, characteristic 105
   species, Mesopotamian Steppe 105
   species, regeneration 108
Climax vegetation, hypothesized 105
Cloud 16
Clover 84, 131
   cultivated 131
   Family 92
   used as fodder 131
Cockscomb Family 82
Coffee 85
Colluvial
   deposits 48
   veneers 38
Colluvium 53
Colonizing species 139
Columbus 83
Commercial extractants 80
Common chicory 80, 82, 85, 119
Common goosefoot 79, 82, 84
Common peganum 80, 97, 110
Common reed 80, 90, 113
Compositae 125
Concrete climatic description 21
Condensation 16, 18
Conduits 35, 61
Cones 94
Conglomerates 31, 35, 108
Coniferous groves, reforested 77
Construction, removing of native flora 140
Convolvulaceae 87
Convolvulus 80, 108
   arvensis 87
   dorycinum 87
Coping mechanisms 21
Coquina 31
Cornus species 125
Cotton thistle 86, 108
Coumarin 92
Crataegus 79
   species 95
Crepis 86, 108
   aspera 86
Crops 108

cereal 113
cereal 84
grain 90
grown 123
irrigated 87
nonirrigated 87
production 31
winter wheat 133
Croton 88
Cruciferae 87
Crucifereae 125
Cucrbitaceae 88
Cucumber Family 88
Cultivars 79
Cultivation 57, 94, 108, 113, 131, 146
   ancient practices 64
   areas 88, 90, 133
   barley 90
   fields 89, 91, 92, 95, 97, 108, 113, 145
   fields, dry farmed 110, 113, 119
   fields, irrigated 110, 113, 119
   fruit trees 105
   ground 89, 92, 93, 96, 133
   land 139
   man's 119, 146
   patterns 110
   practices 101
   quantitative 119
   removing of native flora 140
   soils 133
   terraces, ancient 77
   year-round 113
Culture
   ancient 123
   setting 125
Cupressaceae 88
Cupressus sempervirens 88
Cyclical averages 13
Cyclonic development 13
Cynodon 80, 82, 107
   dactylon 89
Cyperaceae 88, 125, 133
Cyperus 113
   longus 88
Cypress 88, 105, 110
   Family 88
Cypressus 80
   sempervirens 105
Cyprus 11

Daisy Family 85
Dams 64
Dandelion greens 87

Date palm 107
Dates 123
Dead Sea 35, 61, 145
   Basin 61
Debris 86, 124
December 18, 90, 93
Deforestation 113
   damage 108
Delphinium 80
   species 95
Demostachya 82, 107
   bipinnata 90
Denudation 45
Deposits
   floodplain 31
   fluvial fan 31
   terrace 31
Desertification 123, 124, 139, 140
   climatic changes 123
   man's role 124
   Palestine 123, 124
Desert-like 139
   conditions 123, 124
Deserts 15, 83, 85, 90, 105, 140
   areas 107
   historical records 124
   human influence 124
   origin and spread 124
   plant 88
   plant community 94
   species 139
   vegetation 15
Detritus, surface 35
Dew 18, 107
   amounts 18
   frequency 18
   measurements 16
   surface coverage and amounts 28
Dewfall 107
Dianthus strictus 84
Diet
   local 79
   man's 131
Dikili-Tash 123
Dilfah 82
Dines pressure tube anemograph 16
Dioscorides 87
Dipsacaceae 88
Diseases, development of 139
Ditches 85, 86, 88, 89, 94, 95, 97
Diuretics 80
Diurnal
   periodicity 18

   range 16
Dock 79, 94, 95
   Family 94
Dogbane Family 82
Domestic animals 131
Domesticated species 79
Dominant climax species 101
Donkeys 108
Downslope 38
   seepage 53
Drainage 77
   basins 69
Drawings 124
Droughts 21, 82, 93, 105, 108
   climate induced 107
   conditions 146
   period 105
   plant species tolerance of 108
   resistant 88
   season 107
Dry
   air 107
   areas 113
   barren hillside 113
   climates 18, 94
   farmed cultivated fields 110, 113, 119
   farmers 18
   hills 92, 93, 107
   periods 92, 105
   places 90, 91, 93, 95, 97
   rocky places 86, 90
   season 21, 64, 69, 86, 95, 108, 110, 113
   sites 119
   soil moisture 105
   steppe 15
   warm weather 13
   wind 15
   year outflow values 69
Dryness 9
   plants, insensitive to 107
Dudevani dew gauge 16
Dust 86
   storms 13, 15, 16
Dwellings 82, 83, 86, 90, 133
Dyes 80
   "turnsole" 89

Early Bronze Age 84
Earth 87
East Africa 11
East Bank settlements 18
Eastern Mediterranean 84

Ecclesiastics, Book of 83
Echinops 108, 113
   species 86
Echium species 83
Ecological
   characteristics, influence on inhabitants 101
   factors 101
   mapping units 119
   niches 77
   principles 145
Ecological units
   boundaries 110
   defined 110
   dominant features 110
   Dry, Barren Hillsides 110, 113
   Dry Farmed Cultivated Fields 110
   flora 110, 113
   Irrigated, Cultivated Fields 110, 113, 119
   Moist wadi floors 110, 113, 119
   patterns 110
   prominent characteristics 110
   secondary 110
   surficial materials 110, 113
   topography 110, 113
   watercourses 113
   zones 101
Ecology 101
Economy 98
Ecosystems, patterns 110
Egypt 93
El Al 145
El Manshiya 35
Elevation 16
Emetics 89
Environment 113
   arid 48
   biotic, ancient 145
   changes 123, 139, 145, 146
   climatic factor 21
   conditions 145
   dynamic climates 21
   effects of man 123, 124
   exposed 108
   forest 146
   grassland 146
   history 48, 123
   indicators 123
   local 79
   natural 146
   parameters 105
   physical 123
   potential 110
   problems 21

   stagnant life modes 21
   steppe type 57
Ephemerals 139
Equatorial low 11
Erodium 79, 82, 89, 133
   species 89, 133
Erosion 25, 28, 35, 45, 107, 124, 146
   fluvial 45
   gully 21, 146
   natural 108
   rates 35
   rill 21
   soil 57, 146
   surface 38
   wadis 38, 107
   wind 21
Erucaria species 87
Eryngium 79, 108, 110
   creticum 97
Erysimum crassipes 87, 88
Esculent 80, 87
Ethnobotanical interpretations 125
Eucalyptus 93
Eucalyptus species 93
   rostrata 93
Euphorbia 80, 89, 107, 108, 110
   species 89
Euphorbiaceae 88
Eurasian landmass 15
Europe 13, 96
Evaporation 11, 16, 70, 92
   rates 16
Evapotranspiration 11, 15, 70
Evening 85
Evergreens 84, 88
Excavation 124
Export 131

Fall 82
Fallow ground 89, 93, 95
Famine 84
Fan 45
Faqqoos el-homaar 89
Farm 69
Farmers 95, 113
Farming 45, 57
Fat content 84
Feather grass 82, 90
Fences 80, 83, 84, 86, 89, 96
Ficus 79, 80, 131
   carica 93
   species 131

Ficus (*Continued*)
   sycomorus 93
Fiddleneck 133
Field bindweed 80, 87
Fields 80, 86, 89, 119, 146
   agricultural 85
   cleared 92
   crops 57, 69
   cultivated 83, 87, 89, 91, 92, 95, 97, 108, 110, 113, 145
   cultivated, dry farmed 110, 113, 119
   cultivated, irrigated 113, 119
   dry 90
   grazed 92
   irrigated 110
   nonirrigated cultivated 87
   okra 88
   plowed 108
   rocky open terraced 89
   sandy 88, 97
   tobacco 88
   tomato 88
   topography 108
   wheat 87
Figs 79, 80, 123,
   cultivation 93
   dried fruit 131
   fodder 131
   groves 95
   imported 131
   irrigated 131
   poultice 93
   sweetener 131
   used as a laxative 93
   wild 93
Figwort 96
   Family 96
Filaree 133
Fis el kilaab 82
Fiss-ul-kilaab 84
Flaking 31
Flash flooding 45
Flint 31
Flints 28
Floodflow 64, 69
Floors
   deodorizing 80
   dirt 92
   house 92
Flora 21, 87, 110, 113, 124
   assumed climax 110
   change 139, 140
   characteristics 77, 101, 110, 119

collection 77
descriptions 77, 79
distribution 110
diversity 77
economic uses 77, 79
ecosystem data 77
environmental parameters 105
habitats 77, 79
identification 77
Iron Age 140
local 80, 110, 119
man's influence 110
modern 124, 125, 139
modified 140
native, removed 140
observed uses 79
past 77, 123, 124
physical characteristics 79
present-day 105, 110, 119, 123
type indicative of disturbed areas 77
variety 79
weed-dominated 140
Flotation 124
Flour 82, 84, 125
Flow volumes 70
Flowers 79, 82, 83, 84, 85, 86, 87, 88, 89, 90, 91, 92, 93, 94, 95, 95, 96, 97
Flurial
   channels 113
Fluvial
   deposits 48, 53
   fan 38
   gravel 108
   terrace 38
Fodder 80, 82, 90, 92, 131
Fog 18
Foliage 113
Food 82, 133
   forage 133
   man 131
   people's quest for 146
   storage use 35
   systems, abatement 146
   systems, intensification 146
Footpaths 133, 145
Forage 110, 113, 131, 139
   clover 131
   grass 90
   livestock 133
   plants 90, 92, 133
   species, removal 140
   value 139
Forests 92, 94, 95, 146

ancient 146
cover 146
gradual removal 146
Fossil content 28
Freezing 18
Frontal precipitation 18
Frost 13, 16, 18, 105
Fruit 82, 83, 84, 85, 87, 88, 89, 93, 94, 95, 96, 97, 123
  edible 79, 93, 94, 95
  nut-like 94
  pod-like 92
  trees 105
  wild 79
Fuel 94, 95, 146
  olive 131
  people's quest for 146
  plants used as 92
Fumaria species 125, 133, 139
Fumariaceae 125
Fumigation 80, 91
Fumitory 133
Funnel flow 38

Gabaat 95
Galingale 88, 113
Gardens
  domestic 77
  kitchen 79, 83
Garique 107
Gemmeiz 93
Geology 107,
  characteristics 25, 101
  data 25
  factors 31
  faulting 31
  history 25
  landscapes 145
  local 35
  patterns 110
  processes 25, 31, 45
  surficial 35
  time span 28
Geomorphic Processes 38
Geraniaceae 89, 125, 133
Geranium Family 89
Germination 110
  temperatures 139
Gezer 123
Ghab 90
Ghassa 91
Glaucium species 94

Globe thistle 86, 108, 113
Glomez 93
Goats 77, 83, 87, 92, 95, 108
Goosefoot 84
Gopher wood 88
Gordaab 94
Grains 123, 125, 139
  crops 90
  growing 35
Gramineae 89, 125
Grapes 77, 123
  cultivated 131
  eaten 131
  wine 131
Grass 38, 89, 91, 110
  Family 89
  forage 90
  grazing 90
  perennial 107
  steppe 90
  lands 146
Gravel 31, 53
  fluvial 108
Gravitation 18
Grazing 21, 35, 45, 57, 87, 90, 97, 108, 146
  animals 77, 95, 113, 146
  areas 113
  damage 108
  heavy, plant toleration of 119
  intensive 105
  man 119
  over 108, 110, 119, 124, 131, 140
  patterns 110
  practices 101, 146
  pressure 108, 119, 146
  quantitative 119
  vegetation disturbance 110
Greens 84, 95
  dandelion 87
  edible 133
Grinding mills 64
Gromwell 133
Ground 87, 94, 97, 146
  cover 21
  cultivated 83, 89, 93, 96, 133
  dry 133
  fallow 89, 95
  fogs 18
  rocky 83, 86
  seed 84
  stony 83, 93
  surface 89
  uncultivated 85, 86, 89, 90, 97

Ground (*Continued*)
   waste 83
Groundwater 61, 69, 70
   accessibility 61, 70
   aquifers 18, 61, 69, 70
   available 70
   calcium 70
   contaminant 70
   hydrology 25
   magnesium 70
   quality 61
   quantities 61, 69
   resources 69, 119
   runoff 64
   sodium 70
   upper reservoir, electrical conductivity 70
   upper reservoir 70
Groves 93
Growth
   conditions 113
   plant 105
Gully 38
   erosion 146
Gum 80, 82
   tree 93
Gypsum 107

Habitat 139
   degraded 108
   dry 133
   favorable 110
   range 140
   vegetation 119
   wet 133
   wildlife 113
Hail 16, 105
Hairy Amaranth 82
Halfa 90, 96
Halib ul bum 89
Halophyte species 107, 133
Hamham 83
Hamlets 145
Hammada, plant community 94
Hand cup anemometers 16
Handab 85
Handakuk 92
Handal 88
Harmal 97
Harvest 95
Hashemite Kingdom of Jordan 9
Hatab 96
Hawa 86

Hawksbeard 86, 108
Hawshez 96
Hawthorne 79, 95
Headed savory 79, 80, 91, 107
Headwaters 61
Heat 93
   extreme 91, 107
   plant species tolerance of 108
   plants, insensitivity 107
   prolonged 105
   regime 15
Hecilar 123
Hedges 80, 87
Heliotrope
   European turnsole 83
   ornamental 81
Heliotropium
   europaem 83
   species 125, 131
Hellenistic period 131, 140
   seeds 124, 133, 139
   taxa 125
Herbaceous 88, 93
Herbs 84, 85, 86, 87, 94, 95, 96, 97, 107, 113
Herds, grazing 146
Hesban, village 86, 131
Hides, dressing of 94
Highlands 38, 45, 53, 57, 61, 95
   plateau 145
Hills 28, 38, 45, 61, 64, 94, 110, 113, 145
   barren 146
   denudation 38, 45, 53
   deteriorated vegetation 108
   dry 85, 93, 107, 110, 113
   eroded 53, 113
   limestone 94, 95, 96
   plateau 146
   reforested 105
   rocky 31, 85, 95
   summits 113
   sunny 92
   wadis 146
Hilly regions 146
Himalayan Range 13
Hindu region 83
Hinterlands 101
Historical
   records 123, 124
   times 123, 145
Hoar frost 18, 21
Holocene 9
Holy thistle 79, 86, 108, 113
Hommaad 94

Honey
  bees 87
  nectar 93
  plant 92
Hordeum 82, 90, 131
  distichon 131
  species 90, 125
  vulgare 125
Houses 80, 88
  building materials 146
  construction 88, 94
Humans 21, 84
  alteration 21, 146
  climate 21
  culture 123
  ecological changes 21
  environment interface 9
  influence, deserts 124
  intrusion 21
  land relationships 16
Humidity
  levels 9, 21, 105, 146
  margins 15
Hydrological
  conditions 146
  cycle 18
Hydrophilic plants 90
Hydropower 64
Hypericaceae 91
Hypericum 80
  crispum 91
  triquetrifolium 91

Ice age 38
Identification of seeds 124
India 83
Indians 82
Indus plain 11
Infiltration 61
Inhabitants
  effect of ecological units 110
  livelihood 101
  way of life 101
Insecticides 80, 83, 95
Intensification 146
Intensive grazing pressure 105
Inversion fog 18
Irano-Anatolian species 87, 88
Irano-Turanian 83
  community 92
  derivative 86
  groups 91

plants 85
territory 85, 101
Iraq 84
Iron 48
Iron Age 140
  flora 140
  occupation 139, 140
  seeds 124
  taxa 125
Iron staining 48
Irrigation, 64, 69, 70, 113, 131, 140
  areas 64, 110
  cultivated fields 113, 119
  divisionary structure 64
  farming 69
  past development 64
  potential 64
  systems 64
Isaiah, Book of 95
Islamic period 139
  seeds 124
  taxa 125
Isohyet 15
Israel 48
  biblical 123

Jalul 87, 97
January 16, 18, 96
Jarmo 123
Jaundice 80, 85
Jericho 123
Jet stream 13
Job, Book of 84
Jordan 13, 15, 16, 21, 25, 61, 84, 87, 89
  climate 82
  Department of Antiquities 101
  government 145
  Hashemite Kingdom 9
  Local Standard Time 15
  National Master Water Plan 61
  Prince Ra'ad 9
  reconnaissance survey 45
  standard time 15
  Valley 145
Jordan River 38, 45, 48, 61, 146
  Basin 61
  Valley 3, 28, 31, 38, 48, 57, 61, 64, 70, 101, 105, 107
  Dead Sea basin 146
Juice 83, 89
Jujube 79, 80, 95

July   18, 82, 83, 85, 86, 87, 90, 91, 92, 94, 95, 96
June   13, 15, 18, 83, 84, 85, 87, 89, 90, 91, 92, 96

Kafour 93
Kahla 83
Khafur 89
Kamsin
  depression 15, 18
  winds 16
Kharma 96
Khamah 91
Khurfaysh ul-jima 86
Knotweed 80, 94, 139
Koddah 88
Koeppen classification system 15
Kurnub
  aquifer 72
  bedrock 72
Kurs'anni 97
Kurunful 84

Labiatae 91
Lachish 123
Lake 123
Lamb's quarters 82, 84, 133
Laminar crust 28
Land 69
  clearing 108
  cultivated 84, 139
  degradation 45
  fallow 93
  forage values 146
  level 69
  ploughing 146
  uncultivated 96, 145
Landforms 35
Landscape 45, 94, 110, 145, 146
Landuse 35, 77, 79, 123
Larkspur 80, 95
Late Roman site 35
Lathyrus sativus 133
Leaves 82, 83, 84, 85, 86, 87, 88, 89, 90, 91, 92, 93, 94, 95, 97, 105, 107
  edible 96
  grazed 105
  summer 91
  used as a pot herb 97
Legumes 140
Leguminosae 92, 125

Lens
  culinaris 125, 131, 140
  species 131
Lentils 123, 131
Lesser reedmace 96
Lethal alkaloid solanin 96
Libya 13
Libyan High pressure cell 13
Lichens 107
Life mode alternatives 21
Liliaceae 92
Lily Family 92, 105
Lime 48, 107
  accumulation 48
  agricultural 35
  kilns 95
  mud 28
  production 28
  soil water 107
Limestone 25, 28, 38, 45, 53, 70, 85, 91, 93, 94, 96, 107
  fossiliferous 31
  hills 94, 95, 96
  natural formations 107
  nonporous, crystalline 28
  quarrying 35
  soils 95
  resistant 28, 31
Lisseq 88
Lithified rocks 28
Lithospermum
  arvense 133, 139
  species 133
Littein 92
Livestock 83
  browse 139
  fodder 131
  forage 133
Loaves 125
Loess 85
Lolium 82, 107
  species 90
  temulentum 90, 139
Long-term adaptations 21
Lovegrass 82, 90, 107
Low pressure cells 13
Low pressure trough 13
Low temperatures 18
Lower Cretaceous age 72
Lower elevations 64
Lower Jordan Basin 11
Loz 95
Lycium 79, 107

species 96

Macroclimate
   conditions 16, 145
   processes 3, 9
   shift 13
Macroprocesses 15, 16
Macrosystems 15
Madaba 16, 18, 61, 84, 145
   Plains 145
Mallow 79, 84, 93
   Family 93
Malt 131
Malva 79, 140
   species 93, 139
Malvaceae 93, 125, 139
Man
   activities 139, 140
   ancient 70
   cultivation 119, 146
   damage inflicted 108
   desertification 124
   diet 131
   domestic activities 35
   early 123
   effect on environment 123, 124
   food 131
   grazing 119
   influence 110, 123
   landuse 123
   modern 70
   occupation 139
   origin and spread of deserts 124
   prehistoric 31
   relationship with environment 123
   removal of forage species 140
Mantle, unfissured 107
Map
   units 101
   units, ecologically defined 110
Maquis 95
   climax 92
   plants 91
March 16, 88, 90, 91, 92, 94, 95, 96
Marine environment rocks 25
Marls 28, 70, 72, 107
Marrubium 80, 113
   vulgare 91
Marshes 88, 131
Masonry manufacture 35
Mats 89, 90
Mauritanian Steppe 85

May 13, 83, 84, 85, 90, 91, 92, 94, 95, 96, 97
Meadows 89, 133
Meddaad 87, 89
Medicago 80, 82
   orbicularis 92
   species 92, 139
Medicinal uses 80, 85, 91
Medicine 80, 82, 92
Medick 80, 82, 92
Mediterranean 13, 45
   area 82
   batha community 91
   countries 94
   cultivated alluvial soils 92
   cypress 80, 88
   depressions 13
   Forest and Maquis Territory 101
   group 91
   oak forests 146
   region 83
   Sea 11, 13
   wild turnip 133
   Woodland Climax Vegetation Zone 101, 105
Mediterrano-Irano-Turanian 83, 86
Melilot 92
Melilotus 80
   species 92
Mentha 79, 80, 113
   microphylla 91
Mesopotamian Steppe Climax Vegetation Zone 101, 105
Meteorological
   observation 9
   variations 21
   Department, Ministry of Communications 15
Microclimate 15, 16, 119
   conditions 16, 21
   processes 9, 15
Microscope 124
Middle East 13, 83, 94
Mignonette Family 95
Milfoil 80, 85
Milk thistle 86
Millstones 31
Minarets 145
Mineral soil 48
Mint 79, 80, 91, 113
   Family 91
   flavor 91
Moisture 18, 90, 105, 108
   absorbtion 107
   collection 107
   conditions 18

162 ENVIRONMENTAL FOUNDATIONS

Moisture (*Continued*)
   conservation of 107
   limited reserves 107
   regime 119
   retained for plants 105
   soil 107
   surface levels 146
Monsoonal low of Asia 11
Montmorillonite 38
Moraceae 93, 125
Morning 85
Morning Glory 87
Mortar 45
Mount Nebo 31, 145
Mountainous, region 83
Mountains 48
   of Iran 13
Mucilaginous substance 93
Mud brick 45
Mulberry Family 93
   fig 79, 80, 93
Mullein 80, 96
Mureybit 123
Mustard 79, 80, 87
   Family 87
Myrtaceae 93
Myrtle Family 93

Nabq 95
Nafal 92
Nahnah 91
Nar 28
Nari 28, 31, 35, 45, 48, 53, 107, 108
   building materials 31
   capping 107
   exposure 35
   formation 28
   genesis 28
   horizons 28
   layers 31
   lower horizon 28
   origins 48
   outcrop 31
   physical characteristics 28
   pockets and crevices 28
   quarried 31
   surface 28
   unfissured 107
   upper horizon 28
   weathering 28
National Water Master Plan for Jordan 61
Natural
   drainage passageways 35
   ecological cycles 21
   ecological systems 21
   environment 5
   fencing 95
   plant communities 119
   solution passages 31
   subterranean chambers 31
   surface water bodies 61
   vegetation 21
Naur 16, 18, 38, 61, 94, 145
Naur-Jerusalem road 95
Near East 84
Nectar 87, 91
Negil 89
Nerium 80, 113
Nerium oleander 82
New World 82, 83
Nightshade Family 96
Nitrogen
   fixation 131
   in soils 92
Noaea 80
   mucronata 84
Nonagricultural plant cover 105
Noncalcareous 48
Noncultivars 139
Noncultivated
   plants 113, 139
   sites 119
Nonpoisonous species 90
Nonselective plant species 119
Normal gravity 15
Normative weather processes 16
North Africa 11, 13
Northern hemisphere 13
November 16, 88
Nubians 92
Nubo-Arabian Shield 25
Nutrients 21, 31
Nuts 79, 96

Oak 38, 94, 105, 110, 119
   forests 146
   species 108
   vegetative cover 108
October 83, 84, 85, 89, 94
Oils 80, 82, 92
   almond 79
   olive 79, 131
Okra 79, 93
   fields 88

Olea 79, 131
  europaea 93, 125, 131
  species 131
Oleaceae 93, 125
Oleander 80, 82, 83, 90, 91, 113
Olives 79, 93, 94, 123, 131
  ancient oil presses 94
  cooking 131
  cultivated 131
  domestic use 94
  eaten 131
  fuel 131
  groves 77, 84, 86, 89, 91, 95
  oil 94, 131
  orchards 92
  pickled 94
  symbol of peace 94
  trees 94
  utilized 131
Olleiq 87
Ononis 85, 95
  natrix 92
Onopordum 108
  species 86
Opuntia 79, 80
  ficus-indica 83
Orach 84
Orchards 105, 145
  areas 105
  year-round cultivation 113
Organic matter 108, 146
Ornamentals 131, 133
Outflow 69
Overgrazing 108, 110, 119, 124, 131, 140

Paleobotany 139
Paleoclimate 9
Paleoecology 124
  interpretations 125
Paleoecosystem 123
Paleoenvironment 123-140
Paleoethnobotany 5, 123-140
Paleosols 48
Paleospecies 139
Palestine 85, 107, 123, 124, 139
  desertification 123, 124
Papaveraceae 94
Papilionaceae 92
Past 119, 123
  flora 124
  temperatures 133
Pastoralists 18

Pasture 85
  natural 92
  plants 80
Paths 77, 91
Peas 123, 131
  Family 92
Pebbles 124
Pedogenic origins 48
Peds 48
Peganum 80, 110
Peganum harmala 97
Pelycopodal biomicrite 31
Pepper tree 80, 82
Peppercorns 82
Perennial
  flow 69
  streamflow 113
Perfume 80, 92
Persian Gulf 11
Peru 82
Petrium 80
PH 53
Phalaris 82, 90, 107
  minor 90
Phlomis 91, 105
  species 91
Phoenix dactylifera 125
Phosphate rich layers 31
Photography 110
Photosynthesis 105, 107
Phragmites 80, 97, 113
  australis 90
  communis 90
  Typha association 88
Physical environment 123
Physiographic Subdivisions
  East Jordanian limestone plateau 35
  highlands 35
  hills 101
  plateau 101
  wadi 101
  Wadi Araba-Dead Sea-Jordan River Depression 35
Piche evaporimeter 16
Pigs 131
Pigweed 79, 82, 119, 133
Pinaceae 94
Pine 105, 107, 110, 119
  Family 94
  species 108
Pink Family 84
Pinus 80, 107
  halepensis 94, 105

Pistachio 79, 80, 94, 105, 110
   cultivated 94
   edible 94
   Family 94
   Irano-Turanian dwarf shrubs 94
   wild 94
Pistacia 79, 80, 105
   species 94
Pistaciaceae 94
Pisum species 131
Pitch 80, 94
Plains 38, 145, 146
Plaited leaved croton 88, 108
Plant communities, 105, 108, 110, 113, 119
   composition 113
   current patterns 110
   local 107, 108, 113
   natural 119
   noncultivated 113
   toleration of drought 108
   wadi floors 108
Plant cover
   natural, clearing of 108
   nonagricultural present-day 105
   present-day 105
Plants, 77-98, 101-119, 123-140
   adaption 107
   aerated, high water table soils 113
   agriculturally important, remains 123
   aromatic 92
   biblical 83
   characteristic 101
   cleared 96
   climax 113, 119
   common 119
   composition 119
   consumption 108
   crop 90
   cultivated 95, 125, 131
   deep rooting 92
   degradation 119
   deodorizing 92
   desiccation 105
   distribution 105
   documented 79
   dominant 95
   dormant 107
   drought resistant 94, 108
   ecology 101
   economic uses 79, 80, 82
   edible 85, 93, 94, 95, 97
   elimination 146
   environmental indicators 123
   eradicated 108
   fixing nitrogen in the soils 92
   fodder 80, 92
   forage 80, 89, 90, 92, 133
   grazing 79, 90, 96, 110, 113, 119
   greens 95
   growth 53, 110, 113, 146
   heat tolerance 108
   herbaceous 88, 93
   honey 92
   hydrophilic 90
   hydrophytic 82
   insensitive to dryness 107
   insensitivity to heat 107
   life 18
   limiting transpiration 105
   local 79, 197, 108
   maquis 91
   marsh 96
   medicinal 83, 97
   moisture retained 105, 107
   natural fencing 95
   nectar 91
   new propagated 107
   nitrogen fixing in the soils 92
   noncultivated 97, 139
   nonselective 119
   nutrition 53, 108
   observed 79
   odor 85, 92
   ornamental 83
   pasture 89
   poisonous 95, 96
   potential composition 119
   potential density 119
   potential distribution patterns 119
   poultice 94
   present 119, 124
   propagation 110
   reproductive cycles 105, 108
   roots 107, 108
   salt tolerating 107
   shallow rooted 113
   shrub-like 93
   steppe 91
   stress 107
   thorny 96, 108, 110, 113, 146
   toleration of grazing 110
   toleration of extreme heat 107
   toleration of prolonged summer drought 105
   transpiration 70
   unpalatable 108, 110, 119, 146
   unpalatable to animals 97

used against snake bite 97
used as a laxative 93
used as a pain reliever 95
used as an insecticide 95
used as blood thinner 92
used as dye 96
used as fuel 92, 94, 95
used for circulatory illnesses 94
used for diarrhea 94
used for dystentery 94
used on boils 93
used on wounds 93
utilized as food 79
utilized for domestic purposes 80
utilized for industrial purposes 80
wasteland 131
water budget 107
water loving 113
water use 70
wild 93, 94, 95
xerophytic steppe 91
Plant-soil-habitat, relationships 77
Plaster 35, 45
Plastic bags 124
Plateau 31, 38, 45, 48, 53, 61, 64, 69, 70, 105, 145
edge 146
hills 113
region 105
remnant, capping rock 28, 38
soils 110
surface 35, 38, 45, 61, 64, 70, 72, 108, 146
Pleistocene 9, 28, 31, 38, 48
Plowing 108, 124
Pods 83, 92
Poisonous species 90
Polar
air 13, 15, 18
front 13
winter 13
Pollen 123
Polygonaceae 94, 125
Polygonum 80, 139, 140
equisetiforme 94, 139
species 139
Polymorphism 85
Ponds 90
Poppy 94
Family 94
Population
biological 145
distribution 101
Porridge 125, 131

Portugal 87
Pot herb 82
Potassium 53
Potential Climax Species 101
Poterium 80, 84, 85, 92, 95, 107
spinosum 95
Poultice 94
Precambrian, times 25
Precipitation 11, 13, 15, 16, 18, 21, 38, 61, 64, 69, 70, 105
conditions 18
dewfall 107
heavy 69
low amounts 21
maximun intensities 69
mean annual 18
plant use 69
regime 15
summer 105, 107
Prehistoric
man 31
times 145, 146
Present-day
flora 105, 119
nonagricultural plant cover 105
seral flora 110
Pressure 16
gradient 11
values 15
Prickly pear cactus 79, 80, 83
Prince Ra'ad, Jordan 9
Productive technology 21
Protein 84
Prunus 79
amygdalus 95
armenica 125
Pterocephalus pulverulentus 88
Pulses 123
Purgative 82, 87, 88, 89
Purslane 84

Qarni morghaat 89
Qataf 84
Qortom 96
Quantitative inferences 110
Quarrying 35
Quercus 105
calliprinos 105

Raceme 95
Radiation 18

Radiation (*Continued*)
  fog 18
  cooling 16, 18
  heating 16
Rain 11, 13, 15, 18, 21, 53, 64, 70, 107, 123, 124, 131, 139, 145, 146
  annual 131, 139
  cloudburst 18
  conservation 69
  gauges 16
  heavy 45, 69
  high 113, 125, 131
  intensities 38
  low 125
  multi-annual average 13
  optimum 139
  requirements 139
Rainwater 146
Rainy season 61, 89, 90
Range 108
  plants 69, 133
Rangeland
  areas 77, 133
  defoliated by overgrazing 124
Ranunculaceae 95
Ravines 82
Recent times, settlement 119
Reed 90
Reedmace Family 96
Reforestation projects 94
Refuse dumping 35
Regional
  survey 3
  variations 85
Relative humidity 11, 15, 16
Repropagation 105
Reseda lutea 95
Resedaceae 95
Reservoirs 61, 64
Resin 80, 88, 94
Rhamnaceae 95
Rheumatism, plants used as treatment of 92
Rhizomes 21, 90, 92
Rhizomous root system 107
River banks 96
River valleys 146
Roads 77, 83, 84, 85, 86, 94
  edges 85, 89
  sides 82, 83, 84, 86, 89, 90, 97, 133
  surface 31
Robeia 91
Rocks 70, 72, 83, 107
  cracks 90
  crevices 91
  dip of the strata 70
  faulting/fracturing 70
  like 28
  outcrops 113
  permeable 72
  surfaces 113
  transmit groundwater 72
  type 70
  water seepage 107
Rocky
  hills 95
  places 85, 91, 92, 95
  slopes 92
Roman
  period, seeds 124, 131
  period, taxa 125
  road, ancient 77
Romans 96
Roots 79, 85, 94, 95, 97, 107, 108
  base 91
  edible 97
  fleshy 93
  rhizome underground matrix 90
  species 79
  systems 85, 107, 124, 131
  tap 86, 96, 108
  used against snake bite 97
Rosaceae 95
Rose Family 95
Rosettes, eaten as a vegetable 97
Rot, resistant 88
Rubus sanctus 96
Rubus sanguineus 96
Ruins 64, 107
Rumex 79, 94
  roseus 94
Runoff 21, 64, 69, 70
  rates 18
Rye grass 139

Sabat 90
Sabbayr 83
Sage 79, 80, 91, 105
Sagebrush 79, 80, 85
Saharo-Sindian flora group 87
Saint John's wort 80, 91
  Family 91
Sakran 83
Salads 79, 87
  components 86
  greens 79

plant 85
Salinity rating 64
Salt 53, 92, 107
Saltwort 79, 82, 84, 107, 108
Salvia 79, 80, 91, 105
   species 91
Sammh 90
Sand 25, 31, 35, 45, 53, 83, 85, 87, 108
   dolomites 107
   fields 88
   places 85, 92, 97
   plains 87
   sediments 25
Sandstone 25, 72, 107, 108
   bedrocks 28, 53
   calcareous 31
   friable 35
   Lower Cretaceous 31, 35
   Pleistocene 31, 35
   Triassic 31, 35
   weakly cemented 31
Sap 82, 83
Sarcophagi 93
Scabious Family 88
Schinus 80
   molle 82
Schrophularia xanthoglossa 96
Scrophulariaceae 96
Scrublands 145
Se'ed 88
Seasons 13
   annual 82, 83, 84, 86, 87, 88, 89, 90, 92, 95, 96, 97
   biennial 85, 92, 95, 96
   deciduous 82, 93, 94, 95, 96
   dry 95, 108, 110, 113
   Evergreen 83, 94
   perennial 83, 84, 85, 86, 87, 88, 89, 90, 91, 92, 93, 94, 95, 96, 97
   spring 84
   averages 13
   variations 16
Seasoning 91, 92
Seawater 107
Sedge 133
   Family 88
Sedimentary rocks 31
Sedimentation 38
Sediments 25, 28, 35, 38, 64
   characteristic landforms 35
   clay rich 38
   discoloration 35
   influence on man's use of land 35
   physical characteristics 35
   processes of deposition 35
   transport 69
Seed producing annuals 105
Seeding 82
Seedlings 87
Seeds 82, 84, 89, 90, 92, 97, 105, 110, 123, 124, 125, 131, 133, 139, 140
   Byzantine period 124
   carbonized 123, 124, 125, 139
   collected 125
   containing alkaloid harmalin 97
   containing alkaloid harmin 97
   damaged 125
   data 124
   dispersion 110
   diuretic derived from 97
   emetic derived from 97
   Hellenistic period 124, 133, 139
   indentification of 124
   Iron Age 124
   Islamic period 124
   modern 124
   Roman period 124
   sample size 125
   site 125
   species 125
   unidentified 139
Semiarid
   areas 28, 123
   climates 15, 16, 31, 61
   conditions 131
Semidesert, region 83
Semisteppe community 92
September 15, 82, 83, 85, 86, 87, 90, 91, 92
Seral community, 92
Sere 140
Sesquioxides 48
Settlements 35, 57
   quantitative 119
   surrounding 101
Settlers 45
Sewage drainage 35
Shade 80, 107
Sha'eer el-faar 90
Sha'eeriyyah 90
Shagaret es-santeen 82
Shajarat ul-felfel 82
Shale 70, 107
Shamal 15
Sharp leaved asparagus 79, 92
Shawk 85, 86, 92
   ud-dardar 85

Shawk (Continued)
  ul jimal 92
  ul-jimal 86
  ul-hanash 84
Sheeh 85
Sheep 77, 87, 92, 108
Shelter 146
Shikuriyyah 85
Shipbuilding 88, 94
Shiqshiq 97
Shobbei 87
Shrubs 79, 82, 83, 84, 85, 87, 89, 91, 92, 95, 96, 107
  adaptation to summer drought 107
  communities 113
  fuel wood 80
  halophytic 84
  Irano-Turanian dwarf 94
  root systems 107
  steppes 84
Sieve 124
Silene cserei 125, 133
Silicic acid 94
Silts 108
Silybum 79, 108, cviii, 113
  marianum 86
Sirocco 15
Sirocco-type winds 15
Sites 107, 108, 110, 123
  disturbed 110
  dry 119
  moist 113
  noncultivated, degradation 119
  wet 107
Siwan 90
Skins, dressing of 94
Sleet 105
Slopes 38, 69, 107, 108, 145, 146
  barren 146
  denudation 38
  denuded 146
  dry 91
  positions 108
Small catchfly 133
Smooth catchfly 133
Snow 16, 18, 105
Society 123
Soil 15, 21, 25, 35, 38, 45, 48, 53, 57, 61, 64, 69, 70, 77, 82, 83, 88, 101, 107, 108, 110, 119
  ability to infiltrate 146
  ability to store rain water 146
  aerated, high water table 113
  after precipitation 105
  aggregates 108
  alkaline 131
  Alluvial 48, 53, 57
  arid climate 57
  azonal 48
  bases 53
  bedrock interface 28
  calcareous 53, 57, 85, 91, 92, 94, 107
  chalky 92
  changes 146
  characteristic 101
  chemical characteristics 31
  chemical composition 53
  chemical properties 45, 48, 53, 57
  classification 45
  clay 87, 92, 108, 110
  coarse textured 113
  compact 53, 108, 113, 146
  conservation practices 57
  consistence 53
  creep 38
  cultivated 57, 108, 113, 133
  depth 53, 70, 108, 110, 113
  developments 25, 48
  drainage 92
  droughty 108
  dry 88, 94, 96, 110
  erosion 28, 146
  evaporation 92
  favorable water retaining properties 108
  fertility 21, 146
  fixing nitrogen 139
  formation 28, 38, 48
  fragment content 57
  genesis 45, 48
  grain structure 53
  gravelly 92
  heavy textured 38
  horizons 28, 48, 53
  horizons, loose surface 108
  horizons, organic matter enriched surface 108
  horizons-A 48
  horizons-B 48, 53
  horizons-C 53
  infiltration 69
  influence on man 45, 57
  influence of man 57
  lime content 53
  limestone 92, 95
  locus 124
  loose 85

modern village 53
moisture 21, 70, 107, 113, 125
Moorman's major classes of 45
nari 108
nitrogen 92
nitrogen fixation 131
noncalcareous 53
nonirrigated 87
nonsaline 53
nutrient content 53, 57, 146
nutrients 110
organic matter 53, 146
organic rich surface 108
Palestinian 38
patent materials 48
patterns 110
physical characteristics 31
physical data 53
physical properties 45, 48, 53, 57
plateau 110
poor 93, 96, 139
rangeland 57
recharge 70
Red Mediterranean 45, xlv, 48, 53, 57
Regosoic 53
Regosolic 48, 53, 57
relic 48
rendzina 89
resources 57
rocky 87, 94, 133
root systems 124
salinity 57, 84
salt 53, 92, 93, 107
sample 124
sandstones 108
sandy 53, 82, 87, 94, 110, 133
shallow 84, 87, 93, 96, 108, 110, 113
stony 108, 110, 113
structure 53
surface 35, 53, 69
surface horizons 146
surveys 45
tell 53
Terra Rosa 45, 89
Terra Rosa type 48
texture 69, 108
thermometers 15
topography 57
under oak vegetative cover 108
waste calcareous 133
water 11, 53, 70, 105, 107, 108, 113
water recharge, annual 105
water retention 57, 69, 108, 110

water seepage from upslope 105
water storage 69, 107
water, underground 108
weathering 53
Yellow 48, 53, 57
Yellow Mediterranean 48, 53, 57
Soil-like 28
Solanaceae 96
Solanum 79
  nigrum 96
Sonchus 79, 80, 108
  oleraceus 86
Sorghum 80, 82, 90
  species 90
South Africa 88
South America 87
Southern Iran 11
Southern Russia 13
Sow thistle 79, 80, 86, 108
Species 105
  biregional 86, 91
  colonizing 139
  description 82-97, 125-139
  desert 139
  diuretics 80
  forage 82
  grain 139
  grazed 80, 82, 90
  household uses 80
  local 80
  medicinal uses 80
  noncultivated 139
  nonpoisonous 90
  observed 79
  past 79
  pluriregional 84
  poisonous 90
  ruderal 82
  segetal 82
  segetal-rudral biregional 83
  suitable for fodder crops 79
  suitable for grazing 79
  unidentified 133
  unpalatable 105
  used as blood thinners 80
  used as chemicals 80
  used as dyes 80
  used as fences 80
  used as fuel 80
  used as hedges 80
  used as mats 80
  used as ornamental trees 80
  used as poultices 80

Species (*Continued*)
   used as remedy for coughs 91
   used as sacred trees 80
   used as shade trees 80
   used as stimulant 91
   used as stimulants or tonics 80
   used as thatching 80
   used for respiratory ailments 80, 91
   used for timber 80
   used medicinally 87, 88, 89, 91
   used to induce urination 80
   used to induce vomiting 80
   used to reduce fever 91
   weed 139, 140
Specimens, domesticated 88
Spices 79
Spinach 84, 96, 133
Spinacia olearcea 133
Spiny Clotbur 87
Springflow 70, 145
Springs 18, 31, 61, 64, 69, 70, 72, 77, 82, 83, 88, 91, 93, 95, 105, 131, 133
   perennial 61, 64, 113
Spruge Family 88
Spurge 80, 89, 107, 108, 110
Standard sample number 124
Star thistle 85, 133
Stellaria 133
   species 133, 139
Stemflow 21
Stems 82, 83, 84, 85, 86, 87, 88, 89, 90, 91, 94, 95, 96
Steppe 15, 85, 105
   grass 90
   plant 91
   type vegetation 15
Steppe-like vegetative cover 146
Stereoscopic aerial photography 64, 110
Stevenson screen 15
Stipa 82, 90
   capensis 90
Stone robbing 35
Stony places 92, 93, 95, 97
Storage
   boxes, manufacture, ancients 88
   facilities 69
Storksbill 79, 82, 89
Storms 64
Strata, resistant 31
Stratosphere 13
Streams 61, 82, 91, 96, 146
   active 97
   banks 88, 89, 90, 110, 131

bedload 45
beds 45, 64, 82, 83, 92, 93, 95
courses 38, 82, 113
edges 90
ephemeral, banks 88
gradients 38
flow 18, 61, 113, 145
perennially flowing 77
permanent, banks 88
Street construction 35
Structures 31
Subhumid
   areas 107
   Palestine 18
Subsoil 48, 53
Subsurface
   flow 108, 113
   water 108, 110, 113
Subtropical
   high 13
   pressure 11
Successional trends 21
Sudanian 83
Sugar content 131
Summer 11, 13, 16, 18, 64, 69, 83, 84, 85, 95, 97, 105, 107
   air 11
   deciduous 107
   drought 77, 105
   irrigated crops 87
   late 91
   precipitation 105, 107
   prolonged drought 105
   season 11, 15
   stability 16
Sun 82, 91, 93, 146
Sunny hills 92
Surface air
   cooling 18
   masses 11
Surface bedrock 28
Surface condensation 18
Surface horizons 146
Surface minimum thermometers 15
Surface moisture levels 146
Surface resources 119
Surface runoff 64
Surface sediments 25
Surface shape 35
Surface soils 28
Surface temperatures 13, 146, 110, 145
Surface water 61
Surface water resources 64

influence of man 69
influence on man 69
natural 69
perennial 61
rates of flow 61
springs 61
water quality 61
Surficial geology 25, 101, 119
  influence of man 45
  influence on man 45
  patterns 110
Surficial materials 25, 38, 45, 53, 107, 110, 113, 119
  characteristic 101
  creep process 38
  depth 110
  history of deposition 25
  physical characteristics 38
  texture 110
Suspended load 69
Swales 38
Swamps 90
Sweet clover 80, 92
Sweet pea 133
Sweetener 131
Switzerland 84
Syrian
  border 35
  Plateau 13

Tamaricaceae 96
Tamarisk 80, 84, 90, 96, 108, 113
  Family 96
Tamarix 80, 108, 113
  species 96
Tannin 80, 94
Tap roots 108
Tares 90
Taxa 125
  Byzantine period 125
  cultivated 125
  Hellenistic period 125
  Iron Age 125
Tea 91
Technology 21
Teen 93
Tel Mashosh 123
Teleilat Ghassul 123
Tell
  El Al 96
  es-Sawwan 84

Hesban 77, 79, 80, 82, 83, 84, 85, 86, 87, 88, 89, 92, 93, 94, 95, 96, 97
  Jalul 145
  Ramad 123
Tell-ed-Duweir 123
Temperature 9, 15, 16, 18, 124, 131, 139
  average 11, 15, 16
  coldest monthly mean 16
  conditions 18
  cool 133
  extremes 16
  germination 139
  high 15, 105
  maximum 15
  measurements 15
  norms 9
  optimum 139
  past 133
  surface 146
Terra Rossas 48
Terrace 45
Terrain 101
Tethys Sea 25
Thatching 90
Thermic
  equator 11
  inversion 11
Thermodynamic
  high pressure 11
  pressure 13
Thermometers 15
Thickets 96, 113
Thistle 80, 85, 86, 108, 113, 140
  cotton 108
  Family 97
  globe 108, 113
  holy 108, 113
  physical characteristics 85
  star 133
Thorns 86, 95
Thorny
  burnet 80, 95, 107
  plants 113
  saltwort 80, 84
Thunder and lightning storms 16
Thyme 91
Thymus 79, 80, 107
  capituatus 91
Tigris-Euphrates lowlands 11
Timber sources 80
Tobacco 88
Tomato 88
Tools 31

Topography 25, 35, 38, 69, 77, 107, 108, 110, 113, 146
  conditions 146
  favorable 113
  feature 28
  high 28
  landform 110
  maps 110
  pattern 110
  shoulder 28
Tracks 86, 95
Tracksides 82, 83, 85, 86, 87, 89, 90, 91, 92, 93, 94, 96, 97
Transitional seasons 16
Transjordan 87
Transjordanian Plateau 3, 28, 31, 45, 57, 61
  climate 105
  surface 61, 105, 110
  western edge 28
Transpiration 84, 91, 105, 107
Trees 79, 80, 82, 84, 88, 93, 94, 95, 96, 105, 145
  citrus 82
  evergreen 88
  fuel wood 80
  on plateau 105
  ornamental 80, 82
  root systems 107
  sacred 80, 95
  shade 80, 82, 95
  source of oil 96
  timber 84, 94
  used for offerings in prayer 95
  young 146
Tribulus 80, 97
  terrestris 97
Tributaries 28, 61, 70
Trifolium 131
  species 125, 131, 139
Trigonella species 92
Triticum 125, 131, 139
  aestivum 125
  species 125
Tropical air 13
Tumbleweed 82, 119, 133
Turnips 133
Turnsole 80, 83, 89
Twigs 82, 92
Typha 79, 90, 113
  angustifolia 96
Typhaceae 96
Typic Camborthids 48

Ul asaf 83
Umbelliferae 97
Umm al-Jamal 35
Uncultivated
  areas 92, 86, 105, 108, 113
  ground 89, 97
  land 96, 145
Underground rhizomous stems 110
United Nations 48
United States Department of Agriculture 45
Unventilated dry bulb thermometer readings 16
Unventilated wet bulb thermometer readings 16
Upper air 11, 13
Upper Cretaceous Carbonate Bedrocks 28
Upper Jordan Valley 9

Valleys 61, 145
Vapor pressure 16
Vegetables 79, 85, 97
  cooked 79
  fields, cultivated 77
  potential sources 79
  year-round cultivation 113
Vegetation 15, 21, 64, 107, 124, 146
  communities 119
  disturbance 110
  environmental factors 77
  habitats 119
  lush 146
  patterns of 77
  present-day deteriorated state of 108
  removal 21
  succession 110
  zonation 101, 119
Vegetation zone
  Mediterranean Woodland Climax 101, 105
  Mesopotamian Steppe Climax 101, 105
Vegatative cover
  cover, oak 108
  cover, potential 119
Verbascum 80
  species 96
Vetches 123
Vicia
  ervilia 131
  faba 125
Villages 53, 145
Vines 87, 105
Vineyards 84, 89, 91, 96
Vipers bugloss 83
Vitaceae 125
Vitis vinifera 125, 131

INDEX

Wadis 38, 45, 53, 61, 64, 69, 70, 72, 101, 105, 107, 108, 110, 113, 131, 145, 146
  Amad 61
  Araba Jordan River Valley 25
  beds 77, 82, 108
  climate 105
  dams 69
  deterorated vegetation 108
  drier areas 48
  erosion 18, 107
  fans 48
  floodplains 48, 53
  floors 45, 108, 110, 113, 119
  floors, plant communities 108
  fluvial fans 53
  Habis 61
  Hesbân 25, 31, 61, 64, 69, 70, 82, 87, 88, 89, 91
  incision 38
  Kafrein 64
  Kanisa 61, 69, 70
  main 146
  Majarr 28
  Manshiya 61
  Muhtariqa 61, 69, 70
  Naur 70, 145
  networks 38
  Nusariyat 61
  ridges 38
  shallow 145
  subsurface flow 108
  subsurface pockets of water 108
  terraces 53
  topography 113
  tracks above the edges 93
  tributary 146
  walls 28
  Zerqa Main 61
Walla Walla College 124
Walls 31, 89, 96
  building 35
  ruins 107
  stone 83
  wooden side 124
Warm front 18
Warm weather 91
Waste
  areas 88
  ground 91
  places 82, 84, 85, 86, 87, 89, 90, 92, 94, 96, 133
Wastelands 133, 139, 140

Water 18, 53, 61, 69, 70, 79, 86, 88, 92, 96, 113, 124
  accumulation 107
  apportionment 69
  availability 113, 113
  characteristic 101
  courses, perennial surface 31
  deficiencies, soil 105
  flotation 124
  flow 72
  hauling 64, 69
  infiltration 53, 70
  irrigation/domestic 69
  obtaining 70
  perennially flowing 57
  plant use 107
  recharges 70
  retention 31, 57
  salinity 64, 70
  seepage 107
  shortages 64
  soil 113
  sources 64
  storage 35, 53, 69, 107
  subsurface 110, 113
  subterranean passage of 31
  supply 18, 64, 70
  surface 31, 110, 124
  table 82, 90
  transport 69
  vapor 18
  well depth 70
Waterbodies 31
Watercourses 38, 61, 64, 80, 113, 146
Water-deficient region 21
Watering, domestic purposes 64
Watering
  place 94
  stock 64
Watersheds 61, 69, 70
Waterwheels 64
Waterworks 64
Wavy Ballota 91, 113
Weapon 31
Weather 87, 91
  conditions 18
  recording 16
Weeds 82, 84, 87, 88, 89, 90, 92, 94, 108, 133, 139, 140
  common 87, 133, 139
  noxious 133
  species 119, 131
Wells 70, 72

DUE DATE